Advances in Manufacturing Technologies and Production Engineering

Edited by

Ilesanmi Afolabi Daniyan

*Department of Industrial Engineering,
Tshwane University of Technology,
Pretoria, South Africa*

Advances in Manufacturing Technologies and Production Engineering

Editor: Ilesanmi Afolabi Daniyan

ISBN (Online): 978-981-5039-77-1

ISBN (Print): 978-981-5039-78-8

ISBN (Paperback): 978-981-5039-79-5

need for a court order if at any point you breach any terms of this License Agreement. In no event will any delay or failure by Bentham Science Publishers in enforcing your compliance with this License Agreement constitute a waiver of any of its rights.

3. You acknowledge that you have read this License Agreement, and agree to be bound by its terms and conditions. To the extent that any other terms and conditions presented on any website of Bentham Science Publishers conflict with, or are inconsistent with, the terms and conditions set out in this License Agreement, you acknowledge that the terms and conditions set out in this License Agreement shall prevail.

Bentham Science Publishers Pte. Ltd.
80 Robinson Road #02-00
Singapore 068898
Singapore
Email: subscriptions@benthamscience.net

BENTHAM SCIENCE

CONTENTS

FOREWORD

With the advent of the elements of the Fourth Industrial Revolution (4IR) and its capability to enhance manufacturing operations, industries are looking for the optimal way to adjust their business models to accommodate some emerging technologies. This is to ensure global competitiveness, increase in productivity, effective response to customers' and market requirements, and reduction in manufacturing lead-time without sacrificing the quality of the final product.

This book brings together some innovative technologies relating to manufacturing and production engineering with a view to assist industries to gain a competitive edge in the face of dynamic production requirements and environments. It also presents some practical guided approaches that could assist manufacturers in making effective decisions, which can translate to sustainability in terms of profitability, material conservation, optimum energy consumption, safety, as well as environmental friendliness. Alongside the expertise of the contributing authors, my teaching and research experiences in production engineering were harnessed to provide recent research, practice, and theories that are relevant to manufacturing, production, and industrial engineering.

Furthermore, the processing of emerging materials for applications in the rail, aerospace, biomedical automobile, and biomedical industries was also considered in this book. Some of the technologies presented in the book are still emerging and will be of great benefit to production engineers, industrial engineers, and manufacturing industries in the quest to solve some of the challenges relating to manufacturing and production activities.

Adefemi, Omowole Adeodu
Department of Mechanical and Industrial Engineering
University of South Africa
Florida, Johannesburg
South Africa

PREFACE

This book titled "Advances in Manufacturing Technologies and Production Engineering" uncovers some cutting edge technologies for product development in response to the demands relating to customers' satisfaction, product quality, productivity, and sustainability. This book offers significant contributions to some emerging manufacturing technologies, and production processes in line with the principles of the Fourth Industrial Revolution.

The book covers both the theoretical and practical concepts of manufacturing technologies and production engineering. Interesting topics relating to materials development, composite development, additive manufacturing technology, Lean assessment tools, manufacturing productivity assessment, industrial performance analysis, predictive analysis, as well as assembly operations were covered in this book.

The book also contains notable research findings on innovative approaches for products development and manufacturing technologies. With the increasing need to ensure sustainability in manufacturing or industrial operations in terms of product development, effective material, cost, energy, andenvironmental sustainability, this book reports on notable findings that could offer a realistic possibility in this regard.

The book disseminates knowledge critical for products development and contributes technologically driven novel ideas for promoting efficiency and quality of manufacturing or production processes with regards to environmental and social impacts.

Ilesanmi Afolabi Daniyan
Department of Industrial Engineering,
Tshwane University of Technology,
Pretoria, South Africa

ACKNOWLEDGEMENTS

I wish to appreciate God Almighty for the grace and privilege to serve as the editor of this book. I wish to thank my precious wife: Oluwatoyin Esther Daniyan, for her invaluable support all the time. I also appreciate God's precious gifts, my dear sons: Daniel and Samuel Daniyan. Many thanks to all the authors who contributed their ideas and findings to this book and to all the reviewers who made time out of their busy schedules to ensure a double blinded review process. My appreciation goes to Dr. Adefemi Adeodu and Dr. Boitumelo Ramatsetse for their professional contributions.

Thank you all.

DEDICATION

This book is dedicated to God Almighty: Most Glorious, Most Magnificent, Immortal, Invisible, and the Only Wise God.

v

List of Contributors

Abiodun Ayodeji Ojetoye	Department of Mechanical Engineering, University of Ibadan, Ibadan, Nigeria
Adefemi Adeodu	Department of Mechanical & Industrial Engineering, University of South Africa, Florida, South Africa
Adekunle Akanni Adeleke	Department of Chemical and Petroleum Engineering, Afe Babalola University Ado Ekiti, Nigeria
Adeyinka Sikirulahi Yusuff	Department of Mechanical and Mechatronic Engineering, Afe Babalola University, Ado-Ekiti, Nigeria
Boitumelo Ramatsetse	Educational Information & Engineering Technology, University of the Witwatersrand, Johannesburg, 2000, South Africa
Elizabeth Imuetiyan Omo-Irabor	Department of Mechanical Engineering, and Mechatronics Engineering, Afe Babalola University, Ado-Ekiti, Nigeria
Felix Ale	National Space Research and Development Agency (NASRDA), P.M.B. 437, Abuja, Nigeria
George C. Ogwara	Department of Mechanical and Mechatronics Engineering, Afe Babalola University, Ado-Ekiti, Nigeria
Ilesanmi Afolabi Daniyan	Department of Industrial Engineering, Tshwane University of Technology, Pretoria 0001, South Africa
Kazeem Aderemi Bello	Department of Mechanical Engineering, Federal University, Oye-Ekiti, Nigeria
Khumbulani Mpofu	Department of Industrial Engineering, Tshwane University of Technology, Pretoria 0001, South Africa
Lanre O. Daniyan	Department of Physics and Electronics, Adekunle Ajasin University, P. M. B. 0001, Akungba Akoko, Ondo State, Nigeria
Lateef Owolabi Mudashiru	Department of Mechanical Engineering, Ladoke Akintola University of Technology, Ogbomoso, Nigeria
Matthew O. Afolabi	Department of Physics and Electronics, Adekunle Ajasin University, P. M. B. 0001, Akungba Akoko, Ondo State, Nigeria
Monisola S. Adewale	Department of Mechanical and Mechatronics Engineering, Afe Babalola University, Ado-Ekiti, Nigeria
Moses Oyesola	Department of Industrial Engineering, Tshwane University of Technology, Pretoria 0001, South Africa
Mukondeleli Grace Kanakana-Katumba	Department of Mechanical and Industrial Engineering, University of South Africa, South Africa
Ntombi Mathe	Laser Enabled Manufacturing, National Laser Centre, Council for Scientific and Industrial Research, Pretoria, South Africa
Olatunde A. Oyelaran	Department of Mechanical Engineering, Federal University, Oye-Ekiti, Nigeria
Osarobo Osamede Ogbeide	Production Engineering Department, University of Benin, Benin City, Nigeria

Peter Pelumi Ikubanni Department of Chemical and Petroleum Engineering, Afe Babalola University, Ado Ekiti, Nigeria

Rendani Maladzhi Department of Mechanical and Industrial Engineering, University of South Africa, South Africa

Rumbidzai Muvunzi Department of Industrial Engineering, Tshwane University of Technology, Pretoria 0001, South Africa

Temitayo Mufutau Azeez Department of Mechanical and Mechatronic Engineering, Afe Babalola University, Ado-Ekiti, Nigeria

Tesleem Babatunde Asafa Department of Mechanical Engineering, Ladoke Akintola University of Technology, Ogbomoso, Nigeria

Titus Kehinde Olaniyi Department of Mechanical Engineering, and Mechatronics Engineering, Afe Babalola University, Ado-Ekiti, Nigeria

CHAPTER 1

Introduction

Ilesanmi Afolabi Daniyan[1,*]

¹ Department of Industrial Engineering, Tshwane University of Technology, Pretoria 0001, South Africa

This book propagates some emerging technologies necessary for products development and sustainability. It also highlights some innovations for enhancing manufacturing or production processes. With the increasing complexities of materials, the quest for smart products and changes in production technologies, there is a need for the development of new materials to meet the service and functional requirements. There is also a need to achieve manufacturing sustainability in terms of manufacturing time and cost effectiveness, energy consumption and environmental friendliness. It is necessary for manufacturers to adjust their business models to incorporate the emerging technologies in response to the dynamic market and customer requirements. Hence, the findings of this book are aligned to some of the emerging technologies that characterise the Fourth Industrial Revolution. The book can help manufacturers and production engineers achieve production goals in a smooth, time and cost effective manner.

Chapter two of this book explores the potentials and the drawbacks of titanium alloy, a potential substitute for steel based materials for use in automobile, aerospace and biomedical, and other engineering fields. Its degree of machinability to meet certain functional requirements was also explored from the literature survey conducted.

Chapter three focuses on composite materials development, specifically tigernut fibres mixed with nanoclay/epoxy polymer composites tailored to automotive applications to mitigate the water absorption challenges of natural fibres.

Chapter four presents the assessment of the microstructure and mechanical properties of as-cast magnesium alloys reinforced with organically extracted zinc and calcium. The conventional consideration for selecting Mg alloy elements is based on their corrosion resistance, good hardness, and strength. However, calcium and zinc were added as alloying elements, and the investigation of the

* **Corresponding Author Ilesanmi Afolabi Daniyan:** Department of Industrial Engineering, Tshwane University of Technology, Pretoria 0001, South Africa; Tel: +27 (064) 5298778; E-mail address: afolabiilesanmi@yahoo.com

effects of the alloying elements on the mechanical properties of the magnesium alloy was carried out.

Chapter five presents one of the digital technologies of the fourth industrial revolution; additive manufacturing. The aim of the chapter is to investigate the surface finish of products manufactured from titanium alloy (Ti6Al4V) powders *via* selective laser melting. The chapter provides an insight into the feasible combination of process parameters that will produce the best surface finish during the selective laser melting of Ti6Al4V powders.

Chapter six provides an insight into the feasible range of process parameters that will enhance the surface finish of products developed using Polyethylene Terephthalate Glycol (PETG) filament. Specifically, in this study, the Fused Filament Fabrication (FFF) of the additive manufacturing technology was employed for the development of radiometer casing. Both the numerical and physical experimentations were carried out, thus leading to the development of a mathematical model for the prediction of the surface roughness of the products produced from the Fused Filament Fabrication.

Chapter seven delves into the possibility of integrating a sensor with a product to enhance condition based and predictive maintenance. This chapter is in line with the growing interest in the use of sensors technology for condition based and predictive maintenance. Other forms of maintenance could be costly with time implications. For instance, in corrective maintenance, where the systems break down before repair, there may be payment for compensation and loss of productive time, making the process cost ineffective. For preventive maintenance carried out at a certain predetermined frequency, the call for maintenance may be excessive and not necessarily required. This chapter demonstrates the concept of condition based and predictive maintenance to achieve a balanced time and cost effectiveness during maintenance operations.

Chapter eight investigates the effect of extrusion variables on the mechanical properties and stress distributions of Aluminum 6063 (Al 6063) produced by the Equal Channel Angular Extrusion (ECAE) approach. Aluminum is a widely used engineering material due to its properties such as strong electrical and thermal conductivity, lightweight, and corrosion resistance, among others. However, efforts are being made to address certain drawbacks of aluminum, such as poor fatigue strength and low heat resistance. Hence, the use of the Equal Channel Angular Extrusion (ECAE) metal forming procedure to address certain limitations of aluminum.

Chapter 9 aims to develop lean assessment tools and techniques for maturity evaluation in a warehouse environment that is mostly used for third-party logistics

(3PL).The objective is to examine the performances of the warehouses in terms of productivity, quality, and employee satisfaction.

Chapter ten employs the Markovian analysis of industrial accident data. This study will serve as a guide to manufacturing company stakeholders on the need to create safety awareness among the workforce.

Chapter eleven seeks to survey the key variables that affect the quality of roofing sheets, ascertain their individual and collective roles in quality control, and employ Statistical Process Control (SPC) for quality control. The purpose of this chapter, therefore, is to sensitise manufacturing firms on the need to adopt good engineering practices in manufacturing and maintenance of production facilities.

The study employs the Kendall Coefficient of Concordance (KCC) and Principal component Analysis (PCA) to investigate the identified factors that influence the production of fibre cement roofing sheets. SPC tools were used to analyse customers' complaints and preferences.

Chapter twelve focuses on the development, sizing of an engine, and simulation of an Unmanned Aerial Vehicle (UAV), including the assembly of the whole propulsion system in the UAV for short range missions. This work provides design data for the development of the UAV; hence, it is envisaged that the outcome of the study will be of immense guide to industries, which specialize in the development of UAV.

Enhancing the Machinability of Titanium Alloy (TI6AL4V): A Comprehensive Review of Literature

Ilesanmi Afolabi Daniyan[1,*], **Adefemi Adeodu**[2], **Khumbulani Mpofu**[1], **Boitumelo Ramatsetse**[3] and **Rumbidzai Muvunzi**[1]

[1] *Department of Industrial Engineering, Tshwane University of Technology, Pretoria 0001, South Africa*

[2] *Department of Mechanical & Industrial Engineering, University of South Africa, Florida, South Africa*

[3] *Educational Information & Engineering Technology, University of the Witwatersrand, Johannesburg, 2000, South Africa*

Abstract: Titanium alloys (Ti-6Al-4V) are alloys, which contain a mixture of titanium and other elements. The alloy boasts excellent mechanical properties such as high toughness, high strength to weight ratio, and good corrosion resistance ability. Its excellent mechanical properties as well as its suitability for high temperature applications, make it fit for many industrial applications. However, titanium alloy has low thermal conductivity, which makes it susceptible to poor machinability and dimensional inaccuracies during machining operations. In this study, a comprehensive review of the literature was carried out in order to identify the various strategies suitable for enhancing the machinability of titanium alloy (Ti-6Al-4V). The findings from the survey indicate that the machinability of titanium to the required surface finish can be enhanced in the following ways: use of effective cooling strategies, process design, optimisation of process parameters, selection of appropriate cutting tool, effective process monitoring and control as well as the selection of the optimum range of process parameters, *etc.* It is envisaged that the findings of this work will assist machinists in their quest to achieve sustainability during the cutting operations of titanium alloy.

Keywords: Machinability, Process design, Surface finish, Sustainability, Titanium alloy.

INTRODUCTION

Titanium alloys have excellent mechanical properties such as high strength to

* **Corresponding Author Ilesanmi Afolabi Daniyan:** Department of Industrial Engineering, Tshwane University of Technology, Pretoria 0001, South Africa; Tel: +27 (064) 5298778; E-mail address: afolabiilesanmi@yahoo.com

weight ratio, good ductility and hardness, as well as excellent corrosion resistance ability. Its application in the automobile, aerospace and biomedical, and other engineering fields have been reported most especially in the areas where special properties such as high strength and low weight are crucial requirements [1 - 4]. The widely reported classes of titanium alloys are in five categories: alpha (α), near alpha type, alpha-beta ($\alpha+\beta$), beta (β), and near beta type. The alpha (α) category is referred to as the hexagonal-closed packed crystalline structure (HCP), while the beta (β) is called the body-centered cubic crystalline structure (BCC) [5]. In order to enhance the mechanical properties of titanium alloy, some alloying elements are usually added. These alloying elements belong to two classes, namely: alpha (α) stabilizers and beta (β) stabilizers. Alpha (α) stabilizers comprise elements, such as aluminum (Al), tin (Sn), Gallium (Ga), Zirconium (Zr), and other interstitial elements such as carbon$^{©}$, oxygen (O), and nitrogen (N) [5]. The alpha (α) stabilizers make the titanium alloy fit for high temperature applications [5].

On the other hand, the beta (β) stabilizers consist of elements such as vanadium (V), molybdenum (Mo), niobium (Nb), and chromium (Cr). They are usually added to reduce the phase temperature. Other alloying elements include iron (Fe), copper (Cu), nickel (Ni), and silicon (Si0 which can be added in order to obtain better mechanical properties such as improved strength and chemical stability as well as improved corrosion resistance and machinability [5].

Comparing the alpha and beta types, the alpha boasts better creep resistance, suitability for cryogenic applications, and high temperature applications. On the other hand, the alpha type boasts better corrosion resistance, better forgeability, work hardening, and cold forming capabilities.

Table **1** summarises the strength of the alpha and beta types of titanium alloys.

Table 1. Comparison analysis of the merits and demerits of the alpha and beta types of titanium alloys [5].

Alpa types		Beta types	
Merits	**Demerits**	**Merits**	**Demerits**
Excellent creep resistance	Lower strength to weight ratio	Higher strength to weight ratio	Costly to formulate
Cryogenic applications	Lower forgeability	Higher forgeability	Surface oxygen contamination
High temperature applications	Lower work hardening rates	Higher work hardening rate and cold formability	Low modulus of elasticity

(Table 1) cont.....

Alpa types		Beta types	
Merits	**Demerits**	**Merits**	**Demerits**
Cost effectiveness	Lower corrosion resistance	Higher corrosion resistance	Lower thermal conductivity
Higher modulus of elasticity	Lower strain rate sensitivity	Higher strain rate sensitivity	Excellent creep resistance
Higher thermal conductivity	Lower plastic and super plastic formability	Higher plastic and super plastic formability	Limited cryogenic applications
Higher temperature strength	Lower heat treatability	Higher heat treatability	Limited high temperature applications
Higher weldability	Lower toughness	Higher toughness	Cost effectiveness
Higher machinability	Lower density	Higher density	lower temperature strength
Higher modulus of elasticity	-	-	lower temperature strength, weldability and machinability

Due to the weaknesses inherent in each of the classes of titanium, a carefully formulated class known as alpha-beta ($\alpha+\beta$) was developed to compensate for the weaknesses [5]. This makes the alpha-beta ($\alpha+\beta$) alloy class find extensive application in the industries. The most common alloy which belongs to this class is the Ti6Al4V. When the amount of alpha (α) exceeds that of beta (β) in the alpha-beta ($\alpha+\beta$) formulation, the resulting alloy is known as near alpha (α) type. Conversely, when the amount of beta (β) exceeds that of alpha (α) in the alpha-beta ($\alpha+\beta$) formulation, the resulting alloy is known as near beta (β) type [5].

One of the special features of titanium alloy, which makes it a potential replacement for other conventional alloys, is the high strength to weight ratio. This implies that titanium alloy can be employed in the development of lightweight components without necessarily sacrificing the strength of the component in meeting its service requirements. The development of lightweight components boasts several advantages, such as economic and environmental sustainability [6, 7]. There exist a direct relationship between the energy consumption of a system and the weight of the system. The more the weight of a system, the more energy consumed and the less sustainable the system is in terms of environmental friendliness. The quest for the development of sustainable manufacturing systems can be achieved with the use of lightweight materials for product development. Through the implementation of sustainable manufacturing measures such as the development of components with lightweight products that are environmentally friendly, energy and resource efficient can be manufactured for use. This will enhance optimal energy usage with minimal environmental consequences [8 - 14]. Research have proven that the high strength and poor thermal conductivity of titanium alloy contribute to its poor machinability [10 - 12].

In the automobile, rail, and aerospace industries, the development of lightweight components can enhance the speed of the developed system. The quest for high speed automobile, rail, and aerospace systems is another factor that has placed a premium demand on lightweight materials such as titanium alloy as a substitute for the existing materials. Kim and Wallington [15] state that the replacement of conventional materials such as iron and steel with lighter materials can bring about a significant reduction in the energy consumption and greenhouse gas (GHG) emissions during the use phase of the final product.

However, despite the excellent mechanical properties of titanium alloy, its machinability, most especially at high temperatures, has been a concern. This is due to its low thermal conductivity, which causes high heat retention in the material, thereby making the manufacturing process less sustainable [16, 17]. This makes titanium alloy to be categorised as "difficult to machine materials," most especially at high temperatures. Furthermore, its low thermal conductivity can also bring about the chemical reactions at elevated temperatures thereby causing the formation of built up edges. The material is prone to adiabatic failure with the development of adiabatic shear bands under a high strain rate machining.

Several challenges have been reported during the machining of titanium alloy at high speed and temperatures, such as tool wear, poor surface finish, low rate of material removal, vibration, chatter, high thermal and pressure loads, spring back, and the development of residual stresses in the work piece material [18, 19]. These challenges can, in turn, affect the sustainability of the machining process as well as the quality and performance of the final product [20, 21].

In a bid to tackle these challenges, many authors have proposed different strategies for sustainable manufacturing such as life cycle assessment, computer aided modelling and simulation, process optimization, development of effective cooling strategies, amongst others [22 - 29].

METHODOLOGY

This study employs the search technique for obtaining the articles reviewed. The contents of the final articles selected were analysed and discussed, and learnings were derived from the articles. The study followed a systematic review process undertaken by Abdulrahaman *et al.* [30]. This involves the identification of data sources, keywords search as well as inclusion and exclusion criteria.

Data Sources

The literature survey was carried out in line with the theme of the study, and the literature search was carried from academic research databases. The academic databases consulted were: Scopus, Directory of Open Access Journals (DOAJ), IEEE Explore, Springer, Science Direct, Emerald, Sage, Web of Science, Taylor & Francis, Directory of Open Access Repository (OpenDOAR), Researchgate, Google Scholar, and Wiley Online Library. Next, the articles downloaded were screened and the most relevant ones were selected numbering 75.

Keywords Search

The technique of keyword search, as proposed by Kitchenham *et al.* [31], was used to obtain relevant literature for the review. The keywords considered include: "Titanium alloy- "Machinability of titanium alloy" "Sustainability of titanium alloy" "Cutting operation of titanium alloy- Ti6Al4V" "Classes of titanium alloy", "Properties of titanium alloy" "Optimisation of cutting parameters of Ti6Al4V" amongst others.

Inclusion and Exclusion Criteria

The inclusion criteria for the selected articles were based on the relevance of the articles to the theme of the study, year of publication, empirical results, as well as the nature of the article (peer-reviewed articles). The total number of articles obtained from the database after the search was 8,046. This was followed by the elimination of unrelated articles, and this brought about a reduction in the number of articles to 3,298. Duplicate and old papers (based on the year of publication) were also eliminated. This brought the number of the articles to 1020, and based on the content synthesis of the articles, a total number of 80 written in the English language were finally selected and reviewed. The framework for the inclusion and exclusion of the articles is presented in Fig. (1).

LITERATURE REVIEW

The review covers some aspects such as surface finish and dimensional accuracy of titanium alloy, modelling and simulation of the cutting process of titanium alloy, cooling strategies for enhancing the cutting operation of titanium alloy, power consumption, and energy requirement during the machining operation of titanium alloy as well as the approaches for enhancing the cutting tool life during the cutting operation.

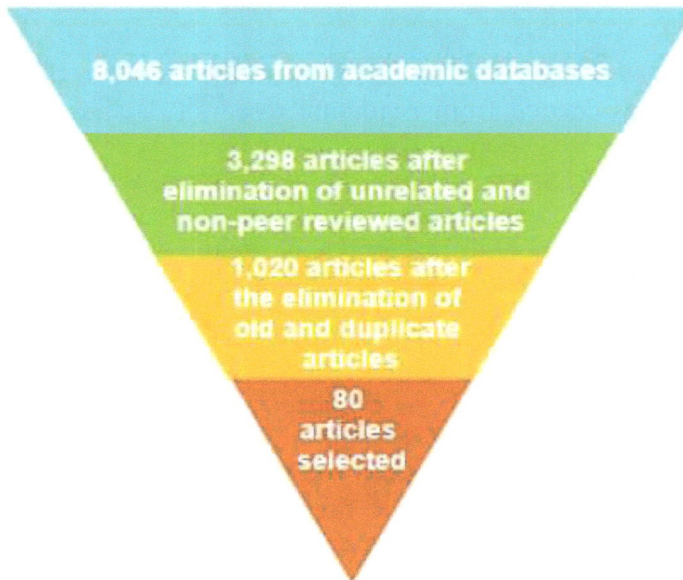

8,046 articles from academic databases

3,298 articles after elimination of unrelated and non-peer reviewed articles

1,020 articles after the elimination of old and duplicate articles

80 articles selected

Fig. (1). The inclusion and exclusion criteria for the articles were obtained.

Enhancing the Surface Finish and Dimensional Accuracy of Titanium Alloy

Mhamdi *et al.* [1] investigated the surface integrity of titanium alloy (Ti6Al4V) during an end milling operation under dry machining conditions. The findings indicate that the orientation of the cutting tool is an important factor that influences the degree of surface finish and the micro hardness of the material. A hemispherical tool at upward and downward milling positions produced the best surface finish surface when compared to machining in the top of the concave surface. The process parameters such as cutting feed, speed *etc.,* were also found to influence the degree of surface finish.

Modelling and Simulation of Titanium Alloy Cutting Operation

In order to enhance the surface finish, sustainability, and machinability of titanium alloy, the use of Design of Experiment (DoE) and mathematical modelling have been reported as viable techniques for achieving a feasible combination of process parameters and for correlating the magnitude of an experimental response such as cutting force, temperature and surface roughness as a function of the independent process parameters [4, 32 - 34].

Cooling Strategies during the Cutting Process of Ti6Al4V

The use of cryogenic MQL cooling during the machining operation of Ti-6Al-4V titanium alloy has been reported [17, 35]. Shokrani *et al.* [35] reported that the cryogenic MQL offered a 30-time increase in tool life during the end milling of Ti-6Al-4V alloy when compared to flood machining. However, Kishawy *et al.* [36] evaluated the sustainability of the machining operation of Ti-6Al-4V using the nano-additives based Minimum Quantity Lubrication (MQL). The sustainability indicators include the environmental, economical, and societal indices. The findings of the study indicate that the cutting operation under the MQL-nanofluid cooling condition offers sustainable performance when compared to that of the classical MQL.

Klocke *et al.* [37] carried out a comparative analysis of the cryogenic cooling (CO_2 & LN_2) and conventional flood cooling during the machining of Ti6Al4V. The results obtained indicate the presence of slight notch wear and uniform flank wear for the machining operation under conventional flood cooling. However, the introduction of cryogenic cooling brought about the reduction in the flank wear and the reduction in the magnitude of the cutting temperature. Bermingham *et al.* [38] reported that the use of high pressure water based emulsion offers improved tool life when compared to cryogenic cooling.

Enhancing the Cutting Process of Titanium Alloy

Arrazola *et al.* [6] carried out a comparative analysis of the machinability of two types of titanium alloy, namely: Ti555.3 and Ti6Al4V. The analysis of the results obtained indicates that Ti555.3 alloy is more difficult to machine when compared to Ti6Al4V alloy. The results further show that Ti6Al4V alloy can be machined at a higher speed compared to Ti555.3. For both materials, the formation of a layer of adhered material composed of Ti and TiC on the rake face surface of the cutting tool was evidenced due to the diffusion process.

Enhancing the Power Consumption and Energy Requirement During the Machining Operation of Titanium Alloy

Tlhabadira *et al.* [39] developed a model for the optimisation of energy consumption during the milling operation of Ti6Al4V. The results obtained indicate a direct relationship between the specific cutting energy and process parameters such as cutting speed and depth of cut but an inverse relationship between the specific cutting energy and the feed rate. The study presented a mathematical model for predicting the cutting energy as a function of the independent process parameters during the milling operation of Ti6Al4V.

Enhancing the Tool Life during the Machining Operation of Titanium Alloy

Elmagrabi *et al.* [40] carried out the investigation of the performance of the coated and uncoated carbide tools under dry slot milling operation of Ti 6Al 4V.

The findings of the study indicate that the Physical Vapour Deposition (PVD) coated carbide tool improved tool life. In addition, the process parameters such as feed rate and depth of cut were found to have significant effects on the surface finish of the workpiece.

CheHaron [41] and Jawaid *et al.* [42] investigated the tool life conditions during the turning operation of Ti-6Al-2Sn-4Zr-6Mo and Ti-6Al-4V under dry machining conditions with the use of uncoated cemented carbide tools. The results obtained indicate that the failure of the cutting tool is attributed to the formation of chips on the flank edge and faces of the tool under increasing cutting speed and feed rates. Hence, the use of cutting inserts with fine grain size and a honed edge were found to increase the useful life of the cutting tool at a feed rate of 0.25 mm/rev.

In order to enhance the useful life of the cutting tool, Ezugwu *et al.* [43] performed the turning operation of Ti-6Al-4V with different coolants using uncoated carbide tools and Cubic Boron Nitride (CBN) tool. The uncoated carbide tools showed better performance compared to the CBN tools in terms of surface finish, failure modes of the tool, and tool wear.

Dandekar *et al.* [44] employed Laser-assisted machining (LAM) and hybrid machining to improve the tool life and the rate of material removal of Ti6Al4V. The machinability of Ti6Al4V improved significantly under the Laser-assisted machining from low to medium (60-107 m/min) cutting speeds, while the hybrid machining improved the machinability of the material from low to high (150–200 m/min) cutting speeds.

DaSilva *et al.* [45] investigated the performance of Polycrystalline Diamond (PCD) tools in turning Ti-6Al-4V at various cutting speeds under wet cutting conditions. The results obtained indicate significant improvement in the tool life with high pressure coolant supplies when compared to the conventional coolant supply. Long continuous chip formation was observed under conventional cooling, while segmented chip formation was observed under high pressure coolant supply.

RESULTS AND DISCUSSION

Findings on Cutting Tool

Ezugwu and Wang [21] explain that the cutting tool for effective machining of titanium alloy should possess the following requirements: high hot hardness to withstand high temperature stress, high thermal conductivity for quick heat dissipation, good chemical inertness to minimise the chances of reacting with titanium, high toughness to withstand chip segmentation and fatigue failure, as well as high tensile, compressive and shear strength. In addition, the authors also emphasise the need for the development of new cutting tools that will meet the service requirement for the machining operations of titanium alloys. The authors, however, recommend the use of cemented carbide (grade WC/Co) for continuous cutting operation and high speed tool for intermittent operation. Pramanik *et al.* [46] explain that ceramic and cubic boron nitride (CBN) tools are prone to large groove wear on the rake face and flank, thus not suitable for machining titanium alloys. However, cutting tools such as carbide, binderless CBN, sintered or natural diamond, are more suitable for machining titanium alloys; their performance depends on the range of parameters used.. Other findings on the cutting tool in respect to various machining operations are captured in Table **2**.

Table 2. Findings on the performance of cutting tool during milling operation of Ti6Al4V.

Author	Cutting Tool	Remarks
Jawaid *et al.* [47]	PVD-TiN and -TiCN/Al$_2$O$_3$ coated carbide tools	The authors found the performance of the CVD coated tool better than the PVD coated tool
Elmagrabi *et al.* [40]	PVD coated carbide tool and uncoated tool	The authors found out that the PVS performed better than the uncoated tool
Wang *et al.* [48]	Binderless cubic boron nitride (BCBN) and Polycrystalline boron nitride (PCBN) tool	The authors found out that the BCBN performed better than the PCBN
Lopez de lacalle *et al.* [49]	TiCN and NCr coated tools	The authors found out that the TiCN coated tools performed better than the NCr coated tools

Findings on Cooling Strategy during Turning Operation of Ti6Al4V

Strano *et al.* [24] explain that the use of cryogenic and conventional cooling of titanium alloy brings about better performance than dry machining. However, cryogenic and conventional cooling might cause cold hardening of the work piece due to low temperature, thus making the material harder and stiffer but less plastic. These cooling conditions bring about an increase in cutting force, reduction in the chances of built up edges, reduction in cutting temperature and frictional activities, improved tool life reduction in temperature, improved surface finish generation of thinner and discontinuous chips [21, 42, 50 - 62]. Fan *et al.* [63] indicate that the cutting of titanium alloy under cryogenic cooling results in improved tool life. The degree of performance of the cutting fluid, however, is a function of the range of process parameters and nature of cutting operations.

Researchers have presented diverging findings on the performance of cryogenic and conventional cooling. Table **3** captures some of their findings.

Table 3. Performance of cryogenic and conventional cooling during the turning operation of Ti6Al4V.

Author	Speed of cutting (m/min)	Feed (mm/rev)	Performance of cryogenic cooling compared to conventional cooling
Venugopal *et al.* [64]	90-117	0.2	Cryogenic cooling improved tool life better than conventional cooling
Dhananchezian & Kumar [52]	27-97	0.159	Cryogenic cooling resulted in the reduction in temperature and cutting force better than conventional cooling
Pusavec & Kopac [54]	60-150	0.254	Cryogenic cooling resulted in improved tool life and reduction in frictional activities better than conventional cooling
Strano *et al.* [24]	50	0.3	cryogenic cooling resulted in the reduction in cutting force and frictional activities better than conventional cooling
Wang & Rajurkar [65]	132	0.2	No difference observed
Bermingham *et al.* [66]	85-125	0.15	Convectional cooling performed better than cryogenic cooling with respect to tool life

Findings on Machining Problems of Ti6Al4V and Possible Remedy

The challenges of machining titanium alloy, causes, and possible remedies are presented in Table **4**.

Table 4. Machining challenges of Ti6Al4V and possible remedy.

Author	Machining Problem of Ti6Al4V	Causes	Remedy
Tlhabadira *et al.* [20], Abele and Frohlich [67]	High thermal stress	Low thermal conductivity	Effective lubrication, Development of better cutting tool materials and coatings to withstand high cutting forces at elevated temperatures, thermally enhanced machining to reduce cutting pressure on the cutting tool surface

(Table 4) cont.....

Author	Machining Problem of Ti6Al4V	Causes	Remedy
Ezugwu *et al.* **[21], Leigh** *et al.* **[68], Sharma** *et al.* **[69],Tlhabadira** *et al.* **[20]**	Friction, vibration, and Chatter	Low modulus of elasticity, deflection under high cutting pressure, spring back	Selection of the cutting tool with the appropriate geometry, use of rigid machines, minimising cutting pressures, provision of copious coolant use of special tools
Su *et al.* **[70]**	Built-up-edges	Ineffective chip removal and cooling strategy	Use of chips breaker and development of effective cooling strategy
Ezugwu and Wang [21],	Wear by diffusion	Ineffective cooling/lubrication, use of the inappropriate cutting tool, friction at the tool-workpiece interface	Effective lubrication, Selection of the cutting tool with the appropriate geometry, the optimum combination of process parameters
Lopez de lacalle *et al.* **[49]**	Hazard of exoergic reaction of Ti chips with atmospheric oxygen	High cutting temperature	Selection of appropriate cutting tool
Lopez de lacalle *et al.* **[49]; Egorova** *et al.* **[71]**	-	Low thermal conductivity	Decreasing the size of the grains and the intragranular microstructural components.
Ezugwu *et al.* **[21], Strano** *et al.* **[24], Zhao** *et al.* **[72]**	Tool wear (Flank wear, crater wear, notch wear)	Built-up-edges, high mechanical loads, high cutting temperature, deflection/spring back, friction at the tool-workpiece interface	Selection of appropriate cutting tool
Bermingham *et al.* **[38], Rashid** *et al.* **[16],Tlhabadira et al [20], Ezugwu** *et al.* **[21]**	Tool fracture	High mechanical loads, high cutting force, thermal and mechanical shock, friction at the tool-workpiece interface, vibration	Coating of cutting tool, selection of cutting tool with the right orientation, redesign of cutting tool
Elshwain *et al.* **[23], Rahman** *et al.* **[73], Sun** *et al.* **[74], Ulutan** *et al.* **[75]**	Poor surface finish	Low thermal conductivity, high temperature stress,	Optimum combination of process parameters, process optimisation, modelling and simulation of the cutting process, use of effective cooling strategy

Findings on the Optimum Process Parameters

For effective material removal with respect to tool life and good surface finish during the machining of Ti6Al4V, Amin [76] recommended cutting speed in the range of 40-80 m/min for uncoated WC-Co inserts and cutting speed within the range of 120-160 m/min using PCD inserts. The use of cutting speeds exceeding this range of values was found to produce chatter, tool wear, and surface roughness. Che-Haron and Jawaid [77] made a similar observation by recommending the use of the uncoated WC-Co at a cutting speed not exceeding 45 mm/min during the machining of titanium alloy. With the use of coated WC tool, Mia *et al.* [78] recommended a higher cutting speed of less than 110 m/min. Ribeiro *et al.* [79] found that the feed rate is a significant factor, which influences the surface roughness and the passive force. It was reported that the depth of cut and feed rate also influences the feed force and cutting force significantly, hence, the need to keep the process parameters within the optimum range. In terms of the corrosion behaviour of Ti6Al4V, Ribeiro *et al.* [79] reported that feed rate and depth of cut are significant factors, which can enhance the reduction in the rate of corrosion of the material when machining is done within the optimum range. Ribeiro *et al.* [79], however, recommended the use of high cutting speed with low feed rate and depth of cut during the machining of Ti6Al4V to obtain a decrease in the passive force, feed force, cutting force, and subsequently surface roughness. This finding agrees significantly with the findings of Oosthuizen *et al.* [80]. Oosthuizen *et al.* [80] reported that a decrease in surface roughness was observed with an increase in the cutting speed and decrease in the feed rate. This finding was attributed to decreasing plastic deformation zone associated with an increase in the magnitude of the cutting speed.

CONCLUSION

The aim of this study was to carry out a literature survey on Ti6Al4V. This is to identify the machining challenges, causes, and possible remedies. Various approaches for enhancing the machinability of titanium alloy (Ti-6Al-4V) were identified in the study. The findings from the survey indicate that the machinability of titanium to the required surface finish can be enhanced in the following ways: use of effective cooling strategies, process design, optimisation of process parameters, selection of appropriate cutting tool, effective process monitoring and control as well as the selection of the optimum range of process parameters, *etc.* It is envisaged that the findings of this work will assist machinists in their quest to achieve sustainability during the cutting operations of titanium alloy.

CONSENT FOR PUBLICATION

Not applicable.

CONFLICT OF INTEREST

The authors declare no conflict of interest, financial or otherwise.

ACKNOWLEDGEMENTS

Declared none.

REFERENCES

[1]   M.B. Mhamdi, M. Boujelbene, E. Bayraktar, and A. Zghal, "Surface integrity of titanium alloy Ti-6A-
        -4V in ball end milling", *Phys. Procedia,* vol. 25, pp. 355-362, 2012.
        [http://dx.doi.org/10.1016/j.phpro.2012.03.096]

[2]   S. Pervaiz, I. Deiab, and B. Darras, "Power consumption and tool wear assessment when machining
        titanium alloys", *Int. J. Precis. Eng. Manuf.,* vol. 14, no. 6, pp. 925-936, 2013.
        [http://dx.doi.org/10.1007/s12541-013-0122-y]

[3]   A. Koohestani, J. Mo, and S. Yang, "Stability prediction of titanium milling with data driven
        reconstruction of phase-space", *Mach. Sci. Technol.,* vol. 18, no. 1, pp. 78-98, 2014.
        [http://dx.doi.org/10.1080/10910344.2014.863638]

[4]   I.A. Daniyan, I. Tlhabadira, S.N. Phokobye, S. Mrausi, K. Mpofu, and L. Masu, "Modelling and
        optimization of the cutting parameters for the milling operation of titanium alloy (Ti6Al4V)",
        *Proceedings of the 2020 IEEE 11th International Conference on Mechanical and Intelligent
        Manufacturing Technologies (ICMIMT 2020),* 2020 Cape Town Added to IEEE Xplore, pp. 68-73,
        2020.

[5]   Y. Oshida, *Materials Classification: In Bioscience and Bioengineering of Titanium Materials* 2ⁿᵈ Ed.
        Elsevier, 2012.

[6]   P.J. Arrazola, A. Garay, L.M. Iriate, M. Armendia, and S. Marya, "L. and Le Maitre, L.
        "Machinability of titanium alloys (Ti6Al4V and Ti555.3)", *J. Mater. Process. Technol.,* vol. 209, pp.
        223-230, 2009.
        [http://dx.doi.org/10.1016/j.jmatprotec.2008.06.020]

[7]   C. Bandapalli, B.M. Sutaria, and D.V. Bhatt, "High speed machining of Ti-alloys- A critical review",
        National Conference on Machines and Mechanisms, vol. 1. 2013, pp. 324-331.

[8]   "Ball, S. Evans, A. Levers and D. Ellison. "Zero carbon manufacturing facility—towards integrating
        material, energy, and waste process flows", *Proc Inst Mech Eng B-J Eng.,* vol. 223, no. 9, pp. 1085-
        1096, 2009.
        [http://dx.doi.org/10.1243/09544054JEM1357]

[9]   K. Branker, J. Jeswiet, and I.Y. Kim, "Greenhouse gases emitted in manufacturing a product—a new
        economic model", *CIRP Ann Manuf Technol.,* vol. 60, no. 1, pp. 53-56, 2011.
        [http://dx.doi.org/10.1016/j.cirp.2011.03.002]

[10]   G. Loglisci, P.C. Priarone, and L. Settineri, "Cutting tool manufacturing: a sustainability perspective",
        The 11ᵗʰ Global Conference on Sustainable Manufacturing, 2013, pp. 252-257 Berlin, Germany.

[11]   H. Lee, D.A. Dornfeld, and H. Jeong, "Mathematical model- based evaluation methodology for
        environmental burden of chemical mechanical planarization process", *Int. J. Precis. Eng. Manuf.
        Green Tech.,* vol. 1, no. 1, pp. 11-15, 2014.

[12]   B. Li, H. Cao, J. Yan, and S. Jafar, "A life cycle approach to characterizing carbon efficiency of
        cutting tools", *Int. J. Adv. Manuf. Technol.,* vol. 93, pp. 3347-3355, 2017.
        [http://dx.doi.org/10.1007/s00170-017-0728-9]

[13]   Q. Yi, C.B. Li, X.L. Zhang, F. Liu, and Y. Tang, "An optimization model of machining process route

for low carbon manufacturing", *Int J. Adv Manuf Technol,* vol. 80, no. 5-8, pp. 1181-1196, 2015.
[http://dx.doi.org/10.1007/s00170-015-7064-8]

[14] K. Gupta, R.F. Laubscher, J.P. Davim, and N.K. Jain, "Recent developments in sustainable manufacturing of gears: a review", *J. Clean. Prod.,* vol. 112, pp. 3320-3330, 2016.
[http://dx.doi.org/10.1016/j.jclepro.2015.09.133]

[15] H.C. Kim, and T.J. Wallington, "Life-cycle energy and greenhouse gas emission benefits of lightweighting in automobiles: review and harmonization", *Environ. Sci. Technol.,* vol. 47, no. 12, pp. 6089-6097, 2013.
[http://dx.doi.org/10.1021/es3042115] [PMID: 23668335]

[16] R.R. Rashid, S. Sun, G. Wang, and M. Dargusch, "An investigation of cutting forces and cutting temperatures during laser-assisted machining of the Ti-6Cr-5Mo-5V-4Al beta titanium alloy", *Int. J. Mach. Tools Manuf.,* vol. 63, pp. 58-69, 2012.
[http://dx.doi.org/10.1016/j.ijmachtools.2012.06.004]

[17] Y. Ayed, and G. Germain, "'Impact of supply conditions of liquid nitrogen on tool wear and surface integrity when machining the Ti-6Al-4V titanium alloy." ", *The Int. J. Adv.Manuf. Tech,* vol. 93, no. 1-4, pp. 1199-1206, 2017.

[18] S. Ehtemam-Haghighi, Y. Liu, G. Cao, and L-C. Zhang, "Influence of Nb on the β→α″ martensitic phase transformation and properties of the newly designed Ti-Fe-Nb alloys", *Mater. Sci. Eng. C,* vol. 60, pp. 503-510, 2016.
[http://dx.doi.org/10.1016/j.msec.2015.11.072] [PMID: 26706557]

[19] A. Pramanik, "Problems and solutions in machining of titanium alloy", *Int. J. of Adv. Manuf. Technol,* vol. 70, pp. 5-8, 2014.
[http://dx.doi.org/10.1007/s00170-013-5326-x]

[20] I. Tlhabadira, I.A. Daniyan, L. Masu, and K. Mpofu, "Computer aided modelling and experimental validation for effective milling operation of titanium alloy (Ti6AlV)", *Procedia CIRP,* vol. 91, pp. 113-120, 2020.
[http://dx.doi.org/10.1016/j.procir.2020.03.098]

[21] E.O. Ezugwu, and Z.M. Wang, "Titanium alloys and their machinability – a review", *J. Mater. Process. Technol.,* vol. 68, pp. 262-274, 1997.
[http://dx.doi.org/10.1016/S0924-0136(96)00030-1]

[22] J.Y. Lee, H.S. Kang, and S.D. Noh, "Simulation-based analysis for sustainability of manufacturing system", *Int. J. Precis. Eng. Manuf.,* vol. 13, pp. 1221-1230, 2012.
[http://dx.doi.org/10.1007/s12541-012-0162-8]

[23] A. Elshwain, N. Redzuan, and N.M. Yusof, "Machinability of nickel and titanium alloys under of gas-based coolant-lubricants (CLS)—A review", *Int. J. Res. Eng. Technol.,* vol. 2, pp. 690-702, 2013.
[http://dx.doi.org/10.15623/ijret.2013.0211106]

[24] M. Strano, E. Chiappini, S. Tirelli, P. Albertelli, and M. Monno, "Comparison of Ti6Al4V machining forces and tool life for cryogenic versus conventional cooling", *Proc. Inst. Mech. Eng., B J. Eng. Manuf.,* vol. 228, pp. 191-202, 2013.
[http://dx.doi.org/10.1177/0954405413486635]

[25] C. Courbon, F. Pusavec, F. Dumont, J. Rech, and J. Kopac, "Tribological behaviour of Ti6Al4V and Inconel 718 under dry and cryogenic conditions–application to the context of machining with carbide tools", *Tribol. Int.,* vol. 66, pp. 72-82, 2013.
[http://dx.doi.org/10.1016/j.triboint.2013.04.010]

[26] G. Rotella, O.W. Dillon Jr, and D. Umbrello, "The effects of cooling conditions on surface integrity in machining of Ti6Al4V alloy." *The Int. J. Adv. Manuf. Tech,* vol. 71, no. 1-4, pp. 47-55, 2014.

[27] K-H. Park, G-D. Yang, M. Suhaimi, D-Y. Lee, T-G. Kim, and D-W. Kim, "The effect of cryogenic cooling and minimum quantity lubrication on end milling of titanium alloy Ti-6Al-4V", *J. Mech. Sci.*

Technol., vol. 29, no. 12, pp. 5121-5126, 2015.
[http://dx.doi.org/10.1007/s12206-015-1110-1]

[28]   G. Kant, and K.S. Sangwan, "Predictive modelling and optimization of machining parameters to minimize surface roughness using artificial neural network coupled with genetic algorithm", *Procedia CIRP,* vol. 31, pp. 453-458, 2015.
[http://dx.doi.org/10.1016/j.procir.2015.03.043]

[29]   H.A. Kishawy, H. Hegab, I. Deiab, and A. Eltaggaz, "Sustainability assessment during machining Ti-6Al-4V with nano-additives-based Minimum Quantity Lubrication (MQL)", *J. Manuf. Mater. Process,* vol. 3, no. 61, pp. 1-12, 2019.

[30]   M.D. Abdulrahaman, N. Faruk, A.A. Oloyede, N.T. Surajudeen-Bakinde, L.A. Olawoyin, O.V. Mejabi, Y.O. Imam-Fulani, A.O. Fahm, and A.L. Azeez, "Multimedia tools in the teaching and learning processes: A systematic review", *Heliyon,* vol. 6, no. 11, 2020.e05312
[http://dx.doi.org/10.1016/j.heliyon.2020.e05312] [PMID: 33195834]

[31]   B. Kitchenham, O.P. Brereton, D. Budgen, M. Turner, J. Bailey, and S. Linkman, "Systematic literature reviews in software engineering–a systematic literature review", *Inf. Softw. Technol.,* vol. 51, no. 1, pp. 7-15, 2009.
[http://dx.doi.org/10.1016/j.infsof.2008.09.009]

[32]   I.A. Daniyan, I. Tlhabadira, K. Mpofu, and A.O. Adeodu, " Development of numerical models for the prediction of temperature and surface roughness during the machining operation of titanium alloy (Ti6Al14V)", *The Int. J. Adv. Manuf. Technol,* vol. 109, no. 7, pp. 1853-1866, 2020.
[http://dx.doi.org/10.14311/AP.2020.60.0369]

[33]   I.A. Daniyan, F. Fameso, F. Ale, K.A. Bello, and I. Tlhabadira, "I. "Modelling, simulation and experimental validation of the milling operation of titanium alloy (Ti6Al4V)", *Int. J. Adv. Manuf. Technol.,* vol. 109, no. 7, pp. 1853-1866, 2020.
[http://dx.doi.org/10.1007/s00170-020-05714-y]

[34]   I.A. Daniyan, I. Tlhabadira, S.N. Phokobye, M. Siviwe, and K. Mpofu, "Modelling and optimization of the cutting forces during Ti6Al4V milling process using the Response Surface Methodology and dynamometer", *MM Sci. J.,* vol. 128, pp. 3353-3363, 2019.
[http://dx.doi.org/10.17973/MMSJ.2019_11_2019093]

[35]   A. Shokrani, I. Al-Samarrai, and S.T. Newman, "Hybrid cryogenic MQL for improving tool life in machining of Ti-6Al-4V titanium alloy", *J. Manuf. Process,* vol. 43, pp. 229-243, 2019.
[http://dx.doi.org/10.1016/j.jmapro.2019.05.006]

[36]   H.A. Kishawy, H. Hegab, I. Deiab, and A. Eltaggaz, "Sustainability assessment during machining Ti-6Al-4V with nano-additives based minimum quantity lubrication", *J. Manuf. Mater. Process,* vol. 3, no. 61, pp. 1-12, 2019.
[http://dx.doi.org/10.3390/jmmp3030061]

[37]   F. Klocke, A. Krame, H.S. Mann, and D. Lung, "Thermo-Mechanical tool load during high performance cutting of hard –to-cut matrials", In: *in Proc.of 5ᵗʰ CIRP Confrence on High performance cutting,* 2012, pp. 295-300.

[38]   M.J. Bermingham, S. Palanisamy, D. Kent, and M.S. Dargusch, "Comparison of cryogenic and high pressure emulsion cooling technologies on tool life and chip morphology in Ti– 6Al–4V cutting", *J. Mater. Process. Technol.,* vol. 212, no. 4, pp. 752-765, 2012.
[http://dx.doi.org/10.1016/j.jmatprotec.2011.10.027]

[39]   I. Tlhabadira, I.A. Daniyan, L. Masu, and K. Mpofu, "K. "development of a model for the optimization of energy consumption during the milling operation of titanium alloy (Ti6Al4V)", *Mater. Today Proc.,* vol. 38, pp. 614-620, 2021.
[http://dx.doi.org/10.1016/j.matpr.2020.03.477]

[40]   N. Elmagrabi, C.H. Che-Hassan, A.G. Jaharah, and F.M. Shuaeib, "High speed milling of Ti 6Al 4V using coated carbide tools", *Eur. J. Sci. Res.,* vol. 22, no. 2, pp. 153-162, 2008.

[41]   "Tool life and surface integrity in turning Tialloy", *J. Mater. Process. Technol.,* vol. 118, pp. 231-237, 2001.
[http://dx.doi.org/10.1016/S0924-0136(01)00926-8]

[42]   A. Jawaid, "C.H. CheHaron and A. Abdullah, A. "Tool wear characteristics in turning of Ti-Alloy Ti-6246", *J. Mater. Process. Technol.,* vol. 92-93, pp. 329-334, 1999.
[http://dx.doi.org/10.1016/S0924-0136(99)00246-0]

[43]   E.O. Ezugwu, R.B. Dasilva, J. Bonney, and A.R. Machado, "Evaluation of the performance of CBN tools when turning Ti-6 Al-4V alloy with high pressure coolant supplies", *Int. J. Mach. Tools Manuf.,* vol. 45, pp. 1009-1014, 2005.
[http://dx.doi.org/10.1016/j.ijmachtools.2004.11.027]

[44]   C.R. Dandekar, Y.C. Shin, and J. Barnes, "Machinability improvement of Ti-alloy Ti-6Al-4V via LAM and hybrid machining", *Int. J. Mach. Tools Manuf.,* vol. 50, pp. 174-182, 2010.
[http://dx.doi.org/10.1016/j.ijmachtools.2009.10.013]

[45]   R.B. DaSilva, A.R. Machado, E.O. Ezugwu, and J. Bonney, "W.F. and Sales. "Tool life and wear mechanisms in high speed machining of Ti-6Al-4V alloy with PCD tools under various coolant pressures", *J. Mater. Process. Technol.,* vol. 213, pp. 1459-1464, 2013.
[http://dx.doi.org/10.1016/j.jmatprotec.2013.03.008]

[46]   A. Pramanik, M.N. Islam, A. Basak, and G. Littlefair, "Machining and tool wear mechanisms during machining titanium alloys", *Adv. Mat. Res.,* vol. 651, pp. 338-343, 2013.
[http://dx.doi.org/10.4028/www.scientific.net/AMR.651.338]

[47]   A. Jawaid, S. Sharif, and S. Koksal, "Evaluation of wear mechanisms of coated carbide tools when face milling Ti-alloy", *J. Mater. Process. Technol.,* vol. 99, pp. 266-274, 2000.
[http://dx.doi.org/10.1016/S0924-0136(99)00438-0]

[48]   Z.G. Wang, M. Rahman, and Y.S. Wong, "Tool wear characteristics of Bindrless CBN tools used in high-speed milling of Ti-alloys", *Journal of Wear,* vol. 58, pp. 752-758, 2005.
[http://dx.doi.org/10.1016/j.wear.2004.09.066]

[49]   "L.N. Lopez de lacalle, J. Perez, J.I. Lorente and J.A. Sanchez. "Advancd cutting conditions for the milling of aeronautical alloys", *J. Mater. Process. Technol.,* vol. 100, pp. 1-11, 2000.
[http://dx.doi.org/10.1016/S0924-0136(99)00372-6]

[50]   S.Y. Hong, Y. Ding, and W. Jeong, "Friction and cutting forces in cryogenic machining of Ti–6Al–4V", *Int. J. Mach. Tools Manuf.,* vol. 41, no. 15, pp. 2271-2285, 2001.
[http://dx.doi.org/10.1016/S0890-6955(01)00029-3]

[51]   S.Y. Hong, and Y. Ding, "Cooling approaches and cutting temperatures in cryogenic machining of Ti–6Al–4V", *Int. J. Mach. Tools Manuf.,* vol. 41, pp. 1417-3147, 2001.
[http://dx.doi.org/10.1016/S0890-6955(01)00026-8]

[52]   M. Dhananchezian, and M. Pradeep Kumar, "Cryogenic turning of the Ti–6Al–4V alloy with modified cutting tool inserts", *Cryogenics,* vol. 51, no. 1, pp. 34-40, 2011.
[http://dx.doi.org/10.1016/j.cryogenics.2010.10.011]

[53]   F. Pusavec, H. Hamdi, J. Kopac, and I. Jawahir, "Surface integrity in cryogenic machining of nickel based alloy-Inconel 718", *J. Mater. Process. Technol.,* vol. 211, pp. 773-783, 2011.
[http://dx.doi.org/10.1016/j.jmatprotec.2010.12.013]

[54]   F. Pusavec, and J. Kopac, "Achieving and implementation of sustainability principles in machining processes", *Adv. Prod. Eng. Manag.,* vol. 24, pp. 151-16, 2009.

[55]   A.K. Nandy, M.C. Gowrishankar, and S. Paul, "Some studies on high pressure cooling in turning of Ti–6Al–4V", *Int. J. Mach. Tools Manuf.,* vol. 49, no. 2, pp. 182-198, 2009.
[http://dx.doi.org/10.1016/j.ijmachtools.2008.08.008]

[56]   Y. Yildiz, and M. Nalbant, "A review of cryogenic cooling in machining processes", *Int. J. Mach.*

Tools Manuf., vol. 48, no. 9, pp. 947-964, 2008.
[http://dx.doi.org/10.1016/j.ijmachtools.2008.01.008]

[57]   K.V.B.S. Kalyan Kumar, and S.K Choudhury, "Investigation of tool wear and cutting force in cryogenic machining using design of experiments", *J Mater Process Tech,* vol. 203, pp. 1-3, 2008. pp. 95–101

[58]   S. Sun, M. Brandt, and M.S. Dargusch, "Machining Ti–6Al– 4V alloy with cryogenic compressed air cooling", *Int. J. Mach. Tools Manuf.,* vol. 50, no. 11, pp. 933-942, 2010.
[http://dx.doi.org/10.1016/j.ijmachtools.2010.08.003]

[59]   S-C. Jun, "Lubrication effect of liquid nitrogen in cryogenic machining friction on the tool-chip interface", *J. Mech. Sci. Technol.,* vol. 19, no. 4, pp. 936-946, 2005.
[http://dx.doi.org/10.1007/BF02919176]

[60]   Y. Su, N. He, L. Li, and X. Li, "An experimental investigation of effects of cooling/lubrication conditions on tool wear in high-speed end milling of Ti-6Al-4V", *Wear,* vol. 261, pp. 760-766, 2006.
[http://dx.doi.org/10.1016/j.wear.2006.01.013]

[61]   L. Li, N. He, and Y. Su, "Effect of cryogenic Minimum Quantity Lubrication (CMQL) on cutting temperature and tool wear in high-speed end milling of titanium alloys", *Appl. Mech. Mater.,* vol. 34, pp. 1816-1821, 2010.

[62]   S. Yuan, L. Yan, W. Liu, and Q. Liu, "Effects of cooling air temperature on cryogenic machining of Ti–6Al– 4V alloy", *J. Mater. Process. Technol.,* vol. 211, pp. 356-362, 2011.
[http://dx.doi.org/10.1016/j.jmatprotec.2010.10.009]

[63]   Y. Fan, M. Zheng, D. Zhang, S. Yang, and M. Cheng, "Static and dynamic characteristic of cutting forces when high efficiency cutting Ti-6Al-4V", *Adv. Mat. Res.,* vol. 305, pp. 122-128, 2011.
[http://dx.doi.org/10.4028/www.scientific.net/AMR.305.122]

[64]   K.A. Venugopal, S. Paul, and A.B. Chattopadhyay, ""Growth of tool wear in turning of Ti-6Al-4V alloy under cryogenic cooling." ", *Wear,* vol. 262, no. 9-10, pp. 1071-1078, 2007.
[http://dx.doi.org/10.1016/j.wear.2006.11.010]

[65]   Z.Y. Wang, and K.P. Rajurkar, "Cryogenic machining of hard-to-cut materials", *Wear,* vol. 239, pp. 168-175, 2000.
[http://dx.doi.org/10.1016/S0043-1648(99)00361-0]

[66]   M.J. Bermingham, S. Planisamy, D. Kent, and M.S. Dargusch, "A comparison of cryogenic and high pressure emulsion cooling technologies on tool life and chip morphology in Ti–6Al–4V cutting", *J. Mater. Process. Technol.,* vol. 212, no. 4, pp. 752-765, 2012.
[http://dx.doi.org/10.1016/j.jmatprotec.2011.10.027]

[67]   E. Abele, and B. Frohlich, "High speed milling of Ti-alloys", *Adv. Prod. Eng. Manag.,* vol. 3, pp. 131-140, 2008.

[68]   E.P. Leigh, J.K. Schuller, and S. Smith, "Advanced machining techniques on Ti rotor parts", *American helicopter Society 56th Annual Forum* Vignia Beach, Virginia

[69]   V.S. Sharma, M. Dogra, and N.M. Suri, "Cooling techniques for improved productivity in turning", *Int. J. Mach. Tools Manuf.,* vol. 49, pp. 435-453, 2009.
[http://dx.doi.org/10.1016/j.ijmachtools.2008.12.010]

[70]   Y. Su, N. He, L. Li, and X.L. Li, "An exprimental investigation of effects of cooling/lubrication condition on tool wear in high-speed end miling of Ti-6Al-4V", *Journal of Wear,* vol. 261, pp. 760-766, 2006.
[http://dx.doi.org/10.1016/j.wear.2006.01.013]

[71]   Y.B. Egorova, A.A Il'in, B.A. Kolachev, and V.K. Nosov. "Effect of the structure in the cutability of titanium alloys", *Metal Sci. Heat Treatment,* vol. 45, pp. 134-139, 2007.

[72]   H. Zhao, G.C. Barber, and Q. Zou, "A study of flank wear in orthogonal cutting with internal cooling",

Wear, vol. 253, pp. 957-962, 2002.
[http://dx.doi.org/10.1016/S0043-1648(02)00248-X]

[73]   M. Rahman, Y.S. Wong, and A.R. Zareena, "Machinability of titanium alloys", *JSME Int. J. Ser. C Mech. Syst. Mach. Elem. Manuf.,* vol. 46, pp. 107-115, 2003.
[http://dx.doi.org/10.1299/jsmec.46.107]

[74]   S. Sun, M. Brandt, and M.S. Dargusch, "Thermally enhanced machining of hard-tomachine materials-a review", *Int. J. Mach. Tools Manuf.,* vol. 50, pp. 663-680, 2010.
[http://dx.doi.org/10.1016/j.ijmachtools.2010.04.008]

[75]   D. Ulutan, and T. Ozel, "Machining induced surface integrity in titanium and nickel alloys: A review", *Int. J. Mach. Tools. Manuf,* vol. 50, pp. 663-680, 2011.
[http://dx.doi.org/10.1016/j.ijmachtools.2010.11.003]

[76]   N.A.K.M. Amin, A.F. Ismail, and N.M.K. Khairusshima, "Effectiveness of uncoated WC–Co and PCD inserts in end milling of titanium alloy—Ti–6Al–4V", *J. Mater. Process. Technol.,* vol. 192/193, pp. 147-158, 2007.
[http://dx.doi.org/10.1016/j.jmatprotec.2007.04.095]

[77]   C.H. Che-Haron, and A. Jawaid, "The effect of machining on surface integrity of titanium alloy Ti–6%Al–4%V", *J. Mater. Process. Technol,* vol. 166, p. 188192, 2005.
[http://dx.doi.org/10.1016/j.jmatprotec.2004.08.012]

[78]   M. Mia, M.A. Khan, and N.R. Dhar, "Study of surface roughness and cutting forces using ANN, RSM, and ANOVA in turning of Ti-6Al-4V under cryogenic jet applied flank and rake faces of WC tool", *Int. J. Manuf. Technol.,* vol. 93, pp. 975-991, 2017.
[http://dx.doi.org/10.1007/s00170-017-0566-9]

[79]   S.L. Ribeiro Filho, and C.H. Lauro, "A.H.S Bueno and Brandon, L. C. Influence of cutting parameters on the surface quality and corrosion behaviour of Ti6Al4V in synthetic body environment (SBF) using Response Surface Method", *Measurement,* vol. 88, pp. 223-237, 2016.
[http://dx.doi.org/10.1016/j.measurement.2016.03.047]

[80]   G.A. Oosthuizen, K. Nnco, P.J.T. Conradie, and D.M. Dimitrov, "The effect of cutting parameters on surface integrity in milling Ti6Al4V", *S. Afr. J. Ind. Eng.,* vol. 27, no. 4, pp. 115-123, 2016.
[http://dx.doi.org/10.7166/27-4-1199]

CHAPTER 3

Development and Characterization of Tigernut Fibres Mixed with Nanoclay/epoxy Polymer Composites

Adefemi O. Adeodu[1,*], **Ilesanmi A. Daniyan**[2], **George C. Ogwara**[3] and **Monisola S. Adewale**[3]

[1] Department of Mechanical and Industrial Engineering, University of South Africa, Florida, South Africa

[2] Department of Industrial Engineering, Tshwane University of Technology, Pretoria, South Africa

[3] Department of Mechanical and Mechatronics Engineering, Afe Babalola University, Ado-Ekiti, Nigeria

Abstract: Natural fibres have gained huge attention from researchers in the field of composite manufacturing in automotive applications due to their low cost, biodegradability, availability, and high performance. However, due to their high hydroxyl content of cellulose, natural fibres are susceptible to water absorption, which invariably affects the mechanical properties of the composite adversely. Researchers have proven that nanomaterials such as Nano silica Carbide (n-Sic) or nanoclay mixed with the polymer composites can overcome the problem. This study investigates the mechanical and microstructural properties of tiger nut fibres reinforced polymer composites tailored to the automotive application. Tiger nut fibres mixed with nanoclay to the size of 50≤µm were used to reinforce epoxy in three levels of loading 2, 4, 6% respectively. Mechanical and microstructural properties of the composites produced were examined. The results showed an increasing trend of 84, 99, 102 MPa and 110, 125, 138 MPa for tiger nut-epoxy and tiger nut-nanoclay epoxy composites, respectively, in the tensile strength. Also, an increasing trend of 80, 84, 88 BHV and 87, 94, 98 BHV were observed for tiger nut-epoxy and tiger nut-nanoclay epoxy composites, respectively, for hardness. Water absorption capacity decreases as the percentage weight fractions of the reinforcement increases but increases as the duration of immersion in boiling water increases. The microstructures showed good interfacial adhesion between reinforcement and polymer matrix when mixed with nanoclay. Tiger nut fibres show a sustainable material useful for automotive applications.

Keywords: Microstructure, Nanoclay, Nanomaterials, Natural fibres, Tigernut fibres.

---

* **Corresponding Author Adefemi O. Adeodu:** Department of Mechanical and Industrial Engineering, University of South Africa, Florida, South Africa; Tel: +27787223718; E-mail: eadeodao@unisa.ac.za

Ilesanmi Afolabi Daniyan (Ed.)
All rights reserved-© 2022 Bentham Science Publishers

INTRODUCTION

Tiger nut (Cyperus Esculentus) fibre is a category of natural fibre (NF) that belongs to the grass family of plant fibres [1 - 5]. The chemical compositions of the natural fibres show that they are naturally rigid and have complicated structures with crystalline cellulose and hemicellulosic matrix microfibril reinforced with amorphous lignin [2, 6]. They contain 60-80% of holocellulose (combination of cellulose and hemicellulose), 5-20% lignin, and moisture content of about 20% as major constituents [6 - 10]. Natural fibres have unique qualities like abundance, non-toxicity, high performance, easy processing at low cost, non-corrosiveness, and their use constitutes no irritation to the respiratory system [2, 4, 6, 11]. More so, the energy consumption by the fibres during processing is about 17% when compared to synthetic fibres like glass fibres [12 - 14]. The rate of processing, recycling, and environmental interaction depends on the use of the composites in different sectors [15]. The use of natural fibres (NFs) have surpassed synthetic fibres (glass fibres) as the preferred reinforcement in polymer composites in recent decades, owing to their environmental and economic benefits. [2]. The search for a solution to the global energy problem as well as environmental sustainability has brought natural fibres into focus as a reinforcing material in polymer composites [2]. Physical and mechanical properties of natural fibre polymer composites (NFPCs) are majorly determined by the cellulose content, micro fibrillar angle, aspect ratio, fibre-matrix adhesion, the volume fraction of the fibres in the composites, fibre dispersion, modulus of the fibre, and toughness of the matrix [7, 16 - 19]. Fibre-matrix adhesion is a very important parameter when it comes to the mechanical properties of natural fibre polymer composites. It determines the fibre-matrix interface which is responsible for stress transfer from the matrix to the fibre [20]. This informs the use of epoxy as the matrix for the development of the NFPCs. Epoxy resin attracted much attention among the thermoset family of materials owing to its excellent properties such as high modulus, low shrinkage in the cure form, good chemical and corrosion resistance as well as acceptable adhesion characteristics. Besides, epoxies can be cured using various ranges of chemicals with different types of curing conditions [8]. Presently, the majority of all the automotive manufacturers using NFPCs to make different components are making use of plant fibres from blast (flax, hemp, kenaf, jute), leaf (sisal, banana), and fruits (cotton, coir). Less attention has been given to grass families like carrots, tiger nut, *etc.* [21 - 23]. Literature is sparse on the industrial applications of tiger nut despite being reported to contain a very high fibre content than most natural fibres (8.91 g/100 g) [24]. This informs the basis of considering tiger nut as a substitute for synthetic fibres in reinforcing polymer in this research. More so, there are some drawbacks associated with the use of NFs for polymer composites [2, 3]. The major drawback of concern is moisture absorption due to the high hydroxyl (OH) content of the hemi-cellulose that absorbs moisture, thereby causing intermittent swelling and shrinking of the NFPCs when wet and dry, respectively [25]. The hydrophilic property of natural fibres affects the mechanical properties of the NFPCs negatively [3, 6, 26 - 30]. According to Ray and Rout [29], water molecules would tend to attract the hydrophilic groups of the fibres and react with the hydroxyl group (OH) of the cellulose to form hydrogen bonds. Fig. (**1**) shows the moisture absorption of NF.

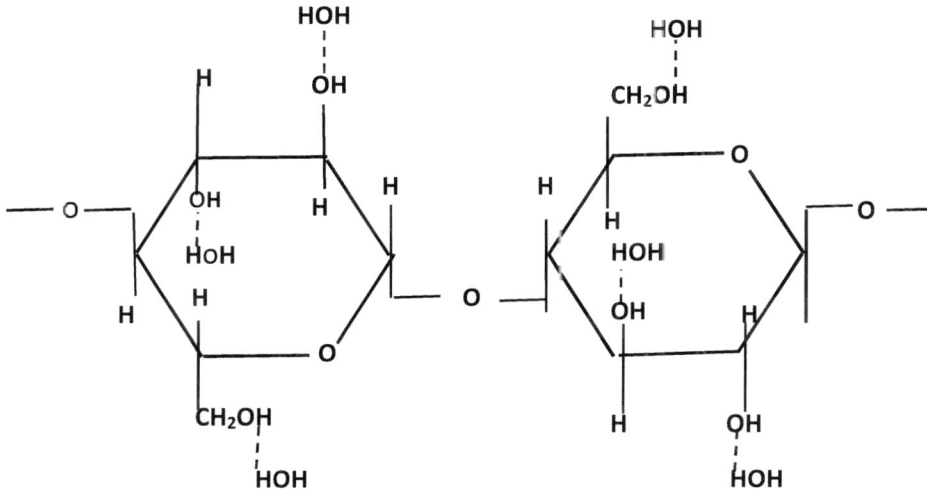

Fig. (1). Moisture absorption of natural fibre [28].

Furthermore, the possibility of the absorbed water forming another layer on the top water molecules previously absorbed may occur. This acts as a separating agent in the fibre-matrix interface leading to separation between fibre and resin [28]. The evaporation of moisture may occur during the process leading to the porosity of the matrix [28]. Fig. (**2**) shows the separation in the fibre-matrix interface.

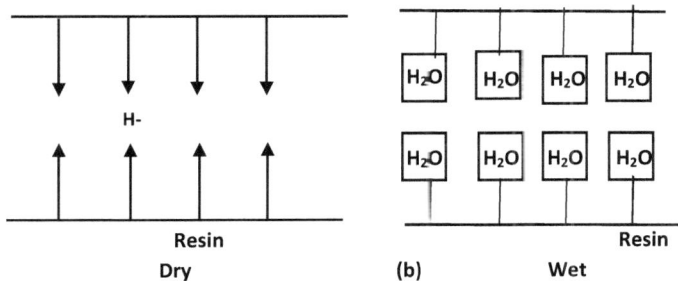

Fig. (2). Separation at the fibre-matrix interface [28].

Research has shown that nanomaterials such as Nano-Silicon carbide (n-Sic) or nanoclay can be used to overcome the problem of water absorption in NFPCs if added to the reinforcement [31]. The addition of nanomaterial enhances mechanical properties like tensile strength, wear, flexural, stress-strain behavior, fracture toughness, and fraction strength of NFPCs both in wet and dry conditions [31]. Nano-Silicon carbide (n-Sic) has received considerable attention in the automotive industry in influencing the physical and mechanical properties of NFPCs compared to nanoclay. As nanotechnology is evolving, clay is increasingly used as natural nanomaterials [32]. Nanoclays are nanoparticles of layered mineral silicates with layered structural units that can form complex clay crystallites by stacking these layers [33]. An individual layer unit is made of octahedral or tetrahedral sheets [34]. The octahedral sheets consist of aluminium or magnesium in six-fold coordination with oxygen from a tetrahedral sheet and with hydroxyl, while the tetrahedral sheet consists of silicon-oxygen tetrahedral linked to neighboring tetrahedral, sharing three corners while the forth corner of each tetrahedral sheet is connected to an adjacent octahedral sheet *via* a covalent bond [35]. The orientation of these sheets displaysthe uniqueness of nanoclays. Most nanoclay stacks are dispersed in a polymer matrix as fillers to form polymer nanoclay composites as they enhance the physical and mechanical properties of the composites [37 - 44]. Based on their chemical compositions and morphology, nanoclays are arranged into various classes such as smectite, chlorite, kaolinite, illite, and holloysite [30]. This informs the rationale behind the use of nanoclay reported in this article. The use of clay in various applications is informed by it wide availability, relatively low-cost, and low environmental impact [31]. More researches have been emerging recently on the development of novel polymer nanoclay composites in the field of material chemistry [36 - 40, 45 - 48]. These novel polymer nanoclay composites are developed with a lower modified nanoclay filler content lighter in weight than conventional polymer composites [48]. The unique properties of PNCCs have made them useful in a number of industrial applications such as automotive [49, 50]. According to Alamri and Low [27], the mechanical properties such as flexural strength increase with increases in the fibre content of the reinforcement. Also, it is observed that the diffusion coefficient increases with an increase in the fibre content. In order to find a solution to the problem of water absorption, Alamri and Low [27] made use of nanosilica carbide mixed with recycled cellulose fibres. It was observed that the modified cellulose fibres decreased the water uptake and enhanced the mechanical properties of the polymer. A similar study was conducted by [69] using nanoclay mixed with sisal fibres as reinforcement. The rate of water uptake was tested on the polymer composites with and without nanoclay. A tremendous decrease in the water uptake and enhanced mechanical properties were observed in the nanoclay filled composites at 5% weight fraction of the reinforcement [69]. According to

Alamri *et al.* [28], the enhancement of the flexural strength of nanoclay recycle cellulose fibres was achieved at 1% weight fraction, while no increment in the property was served at further loading of the reinforcement due to increased voids, viscosity, and poor dispersion. It was also observed that epoxy-recycled cellulose fibre increases flexural strength greatly compared to neat epoxy. Comparing the effect of enhancement of flexural strength by epoxy-nanoclay recycled cellulose fibre and epoxy-recycled cellulose fibre, the addition of the nanoclay is said to be insignificant. Research by Mohan and Kanny [69] shows that nanoclay in the reinforcement changes the stress patterns, thus, contributing to an increase in tensile strength as the weight fraction of nanoclay increases. The nanoclay in the composites changes the failure pattern from brittle to ductile because it acts as a crack arrestor during the loading by inducing deformation mechanisms [69]. Some researchers also used property modification *via* the treatment of the natural fibres with acid or alkalis [70]. Strong alkalis solution can reduce the strength [70]. Therefore, another method of overcoming the drawback of natural fibres to enhance the mechanical properties is property modification. Mohanty *et al.* [8] discussed the enhancement of the mechanical properties of coir-polyester composites in terms of flexural and impact strength *via* alkali treatment. The untreated coir fibre-polyester composites perform lower than the treated. Mohanty *et al.* [8] also studied the chemical surface modification of jute-polyester composites to investigate the performance of the composites. It was observed that mechanical property in terms of tensile strength was improved due to the surface modification.

NFPCs have found considerable applications in the field of automotive and construction due to good specific strength, low density, biodegradability, better tool life, and renewability [51]. Natural fibre polymer composite offers extensive applications in low-cost housing, consumer goods, civil structures, and other different applications [21, 52]. More so, NFPCs were found suitable for paneling elements in airplanes and trains [53]. The global trend in the automotive sector in the direction of lightweight materials coupled with stringent environmental policies have informed the increasing use of NFs [54]. Amongst the numerous natural fibres, only a few of them find good applications in the automotive industries [2]. The use of NFs in automotive industries has widen the scope of research in the field of NFPCs. According to Karus and Kamp [55] and Silver *et al.* [56], the use of polyester and polypropylene (PP) based NFCs for the production of different automobile component parts by the German automotive industries is really gaining attention. Full usage of NFs as reinforcement for door panel, rear panel, seat cover, and various damping and insulation parts was reported by Schih [57]. Al-Qureshi [58] has shown that the use of banana fibre as reinforcement with epoxy resin has excellent fibre-matrix bonding with no delamination. According to Davoode *et al.* [59], kenaf/glass fibre –epoxy hybride composite was reported as structural components in car bumper beams with enhanced mechanical properties. The hybride composites' modulus and tensile strength were superior to those of conventional car bumper beam materials, paving the way for their application as structural components in various car bumper beams.Alves *et al.* [60] used jute fibres as a replacement for glass fibres in the frontal bonnet of a vehicle to investigate environmental, social, technical, and economic sustainability. Jute fibre was found to be better in reinforcement application in all aspects except in the area of technology. A study conducted on the locomotive trains by Indian Railway Industry revealed that NFPCs have significant advantages on various train components like roof panels, berths, modular toilets, *etc.* The use of NFPCs is critical in achieving high speed, low power consumption, weight reduction, less inertia, and low track wear [61].

This study aims to investigate the mechanical and microstructural properties of tiger nut fibres mixed with nanoclay/epoxy polymer composites tailored to the automotive industry with the objectives of determining the effect of nanoclay in the enhancement of tensile, hardness, and morphology of the NFPCs. The use of tiger nut fibres mixed with nanoclay/epoxy polymer composites for automotive applications has not been widely reported by the existing literature.

MATERIALS AND METHOD

The materials utilized in the study are Tiger nut Fibres (TNFs) and nanoclay mixed with epoxy resin as a based matrix. Tiger nut fibres were locally sourced while the nanoclay was supplied by Nanografi Nanotechnology Co. Ltd, Germany. The epoxy resin used in this study was diglycidyl ether of bisphenol A (DGEBA) with an epoxide equivalent mass of 187 g equiv^{-1}. The hardener was XB 3473, which contains diethyltoluenediamine (80-92% concentration) and 1, 2-diaminocyclohexane (4-10% concentration); both were supplied by Huntsman Advanced Materials (Fig. **3a** and **3b**), respectively). All samples were prepared using a stoichiometric ratio of DGEBA-DDM. Polyvinyl Chloride (PVC) was employed as a mold release agent.

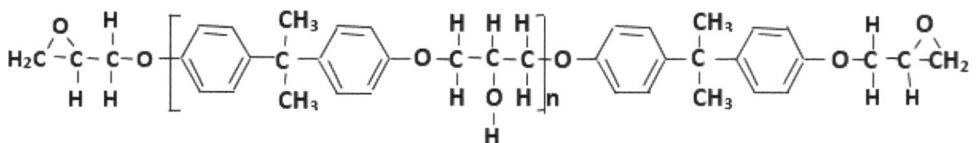

Fig. (3a). Chemical Structure of DGEBA [62].

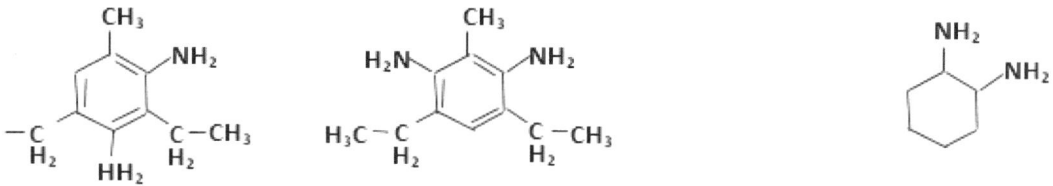

Fig. (3b). Components of the Hardener (diethyltoluenediamine and 1, 2-diaminocyclohexane) [62].

Preparation of Tiger nut Fibres

The tiger nut seeds were first washed to remove the dirt. The juice was extracted with a juice extractor leaving the fibres as by product. The fibres were then sun dried for some days to reduce the moisture content significantly. The dried fibres were milled using a mechanical grounder and sieved into sizes. The two sizes obtained were fine fibres (less than 50 µm) and coarse fibres (100-150 µm). The proximate composition of tiger nut fibres in 100 grams of tiger nut seeds is presented in Table 1.

Table 1. Proximate composition of tiger nut fibres per 100 g of tiger nut tubers.

Moisture	Lipid	Fibre	Protein	Ash	Sugar	Carbohydrate
26.00	24.49	8.91	5.04	1.70	15.42	43.30

Preparation of Epoxy-Tiger nut Fibres Composites

The functionalisation of the tiger nut fibres and nanoclay material was first carried out using microwave heating in order to remove any contaminant in the course of preparation of the reinforcement. Specific masses of TNFs and nanoclay were measured and then mixed with a specific mass of the epoxy-hardener mixture of weight ratio 3:1. The mixing was done using a high speed mixer at 600 rpm for 15 min. Sonification of the mixture was done after the mixing for 30 min using sonicator UP 400S Hielscher: 0.5s cycles with a power of 400 W and amplitude of 50%.

Lamination of the Tiger nut Fibre Polymer Composites and Curing

The epoxy-TNF/nanoclay mixture was used to prepare 200 g of each of TNFs polymer nanoclay composites at different reinforcement loadings of 2, 4, and 6% weighted fraction of TNF/nanoclay. The prepared 200 g samples of different loading of reinforcement were then poured into numbers of 30 x 30 x 30 mm rectangular mould already rubbed with mould releasing agent and then allowed to cure at ambient temperature for 24 hr and finally post cured using the conventional autoclave oven. The temperature of the samples inside the oven was raised up to 140 °C at a heating rate of 3 °C/min for a total duration of 10 hr. The oven was allowed to cool to room temperature before the samples were taken out. A neat epoxy sample of 200 g was also prepared to serve as a control sample. These procedures were repeated for epoxy-TNF only at the same specified percentage weight fraction. The masses of the base matrix and percentage weight fractions of the reinforcements are described in Table **2**.

Table 2. Mixing Ratio of the TNFs Polymer Nanoclay Composite Samples.

Sample	Percentage fraction of TNFs (%)	Mass fraction of epoxy (g)	Mass fraction of hardener (g)	Mass fraction of TNFs (g)	Mass fraction of nanoclay (g)	Mass of nano-composite (g)
1	2.00	147.00	49.00	2.00	2.00	200
2	4.00	144.00	48.00	4.00	4.00	200
3	6.00	141.00	47.00	6.00	6.00	200
4	2.00	147.00	49.00	4.00	-	200
5	4.00	144.00	48.00	8.00	-	200
6	6.00	141.00	47.00	12.00	-	200
7	Neat Epoxy	200	-	-	-	200

The samples of the tiger nut fibre nanoclay polymer composite were machined after pre and post curing to sizes according to the required mechanical analysis carried out in accordance with the ASTM standards (Fig. **4**).

Fig. (4). Tiger nut fibre nanoclay polymer composite.

Tensile Test Analysis

Seven epoxy-TNFs Nano-composite samples were machined and labelled (1, 2, 3, 4, 5, 6, and C) for the tensile test. A. universal Instron 3369 Tensile Machine (TM) was used for the analysis in accordance with ASTM E8/E8M-13a. The original length and breadth of the specimen were measured. The specimen was inserted into the test machine's grips connected with a strain measuring sensor, and a desktop computer was connected to the machine. The load (1 N) was applied with the crosshead speed set 0.5 mm/min until the specimen failed under the loading condition. The record of the tensile strength was automatically generated by the computer connected to the Instron TM.

Hardness Analysis

The hardness analysis was carried out using Brinell's hardness testing machine with 1. 5 mm diameter ball and 100 kg load. This is to measure the extent of a localized penetration by a standardized round or pointed indenter in accordance with the ASTM E384 standard. The specimens were placed one after the other on the hardness testing table and gradually elevated for the specimen to make contact with the indenter. The diameter of the indentation was measured using the microscope and micrometer. The expression for Brinell's hardness number is given in Equation (1).

$$BHN = \frac{2P}{\pi D(D\sqrt{D^2 - d^2})} \tag{1}$$

Where D is the diameter of the indenter in (mm), d is the indent diameter in (mm), and P is the load applied in (N).

Water Uptake Measurement

Water uptake measurement determines the relative rate of absorption of water when the sample is immersed and the weight percentage of water absorbed when saturated. The results provide a guide to the proportion of water absorbed by the material and the effect of the moisture on the mechanical properties of the final product. The analysis was carried out in accordance with ASTM D 570. An analytical balance capable of reading 0.001 g and an oven capable of maintaining the uniform temperature of about 110 °C were used for the test. The test samples were disk with 50.8 mm in diameter and 3.2 mm in thickness with a permissible variation of ± 0.3 mm. The surface of the specimens was polished with 1200 abrasive paper. Three specimens for each loading were prepared. The oven containing distilled water was heated to 80 °C, then the first set of specimens from each loading were immersed in the boiling water for 4 hr, the second set of samples were immersed for 1 day while the third set of samples were immersed for 7 days. The corresponding weight of the samples was measured when removed from the boiling water after 3 min. The difference between the substantially saturated weight and the dry weight was considered as the water absorbed.

Scanning Electron Microscopy (SEM)

SEM study was carried out to examine the change in microstructures due to the addition of TNFs and nanoclay using the electron microscope JEOL-JSM 5800. The samples were prepared and positioned on a sample holder with silver paint and coated with gold to prevent charge build-up by the electron absorbed by the specimens. A 15 kV accelerating voltage was applied to achieve desired magnification [63, 64].

The results of the influence of tiger nut fibres mixed with nanoclay on mechanical and microstructural properties of the natural fibre polymer composites were discussed in this section.

Tensile Test Result

Fig. (5) shows the effect of the percentage weight fractions of tiger nut fibres on the tensile strength and elongation of tiger nut fibre polymer composites. Fig. (4) also presents the effect of nanoclay to the tiger nut fibres (reinforcement) on the enhancement of the tensile strength and elongation properties of the natural fibre polymer composites. From Fig. (5) it is shown that there are increasing trends in the tensile strengths and elongations of the tiger nut fibre polymer composites and tiger nut fibre nanoclay polymer composites as the percentage weight fractions of the reinforcement increases. The increasing trends in both cases were because of the viscoelastic nature of the natural fibres (tiger nut fibres) due to high cellulose content [65] and also an improved ductile behavior as the percentage weight fractions of the reinforcement increases [65]. According to Reddy and Yang [66], tensile strength and elongation of natural fibres are factors such as cellulose content, the ratio of length to the diameter of the fibre, and microfibrillar angle. Comparing the tiger nut fibre polymer composites and tiger nut fibre-nanoclay polymer composites, it is observed that tiger nut fibre-nanoclay polymer composites exhibited higher tensile strength and elongation. This is due to the fact that nanoclay acts as a compatibilising agent by increasing the interfacial bonding between the reinforcement and the polymer matrix, thereby enhancing the tensile strength of tiger nut fibre-nanoclay polymer composites [67, 68]. More so, according to Mohan and Kanny [69], nanoclay acts as a crack arrestor during loading by inducing a deformation mechanism, resulting in the samples failing under the deformed condition. The increase in nanoclay contents changes the stress-strain pattern by increasing the modulus of the composite [3, 4]. Due to the decrease in the water uptake between the tiger nut fibres and the polymer, the equilibrium water content of the composites continuously decreases as the nanoclay content increases. This is due to the fact that nanoclay layers create an impermeable medium to arrest the water flow; hence causing the water molecules to take an indirect path, thus needing more time for water uptake [69]. Table **3** shows the tensile strength and elongation of the composites for varying weight fractions of the reinforcements.

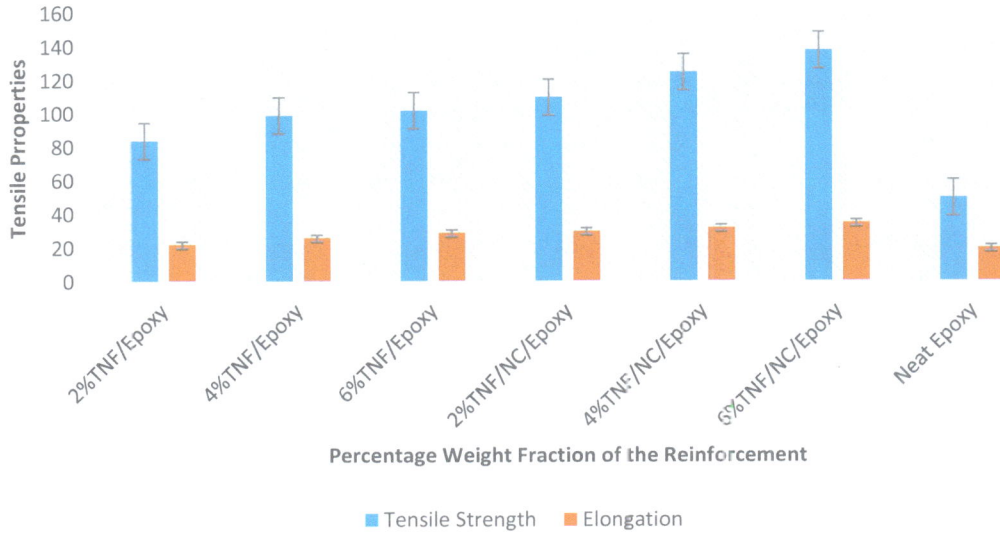

Fig. (5). Variation of tensile strength and elongation against percentage weight fractions of reinforcement.

Table 3. Tensile Strength and Elongation of the Composites.

Samples	2% TNF-Epoxy	4% TNF-Epoxy	6% TNF-Epoxy	2% TNF/NC-Epoxy	4% TNF/NC-Epoxy	6% TNF/NC-Epoxy	Neat
Tensile Strength (MPa)	84	99	102	110	125	138	50
Elongation (mm)	22	26	29	30	32	35	20

Hardness Test Result

Fig. (**6**) shows the variation of Brinell's hardness numbers of the tiger nut fibre polymer composites and tiger nut fibre-nanoclay polymer composites with an increase in percentage weight fractions of the reinforcements. There is a continuous increase in the hardness number as the percentage weight fractions of the reinforcements increases. This can be traced to the volume and modulus of the fibres. Fibres with high modulus tend to impact positively on the hardness of the composites [70, 71]. According to Webo *et al.* [72], the increase in the hardness of the tiger nut fibre polymer composites and tiger nut fibre-nanoclay polymer composites is due to continuous even distribution of the reinforcing materials and corresponding increase in the stiffness of the natural fibre polymer composites. Comparing tiger nut fibre polymer composites to tiger nut-nanoclay polymer composites, the values of the hardness in tiger nut fibre-nanoclay polymer composites are much higher because nanoclay act as the compatibilising agent, thereby increasing the interfacial bonding between the tiger nut fibre and epoxy by improving the fibre-matrix adhesion, thus increasing the hardness [73]. Table **4** shows the hardness of the composites for varying weight fractions of the reinforcements.

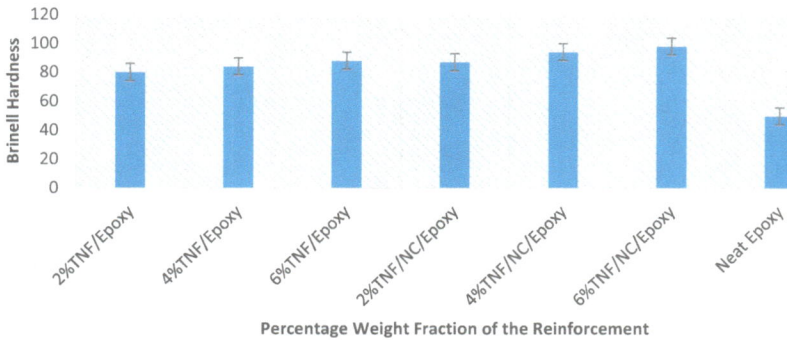

Fig. (6). Variation of hardness to percentage weight fraction of reinforcement.

Table 4. Hardness of the composites.

Samples	2% TNF-Epoxy	4% TNF-Epoxy	6% TNF-Epoxy	2% TNF/NC-Epoxy	4% TNF/NC-Epoxy	6% TNF/NC-Epoxy	Neat
Hardness	80	84	88	87	94	98	50

Water Uptake Result

Fig. (**7**) shows the variations in the percentage moisture content of the tiger nut fibre-epoxy composites and tiger nut fibre-nanoclay epoxy composites as the percentage weight fractions of the reinforcement increase. It is observed from Fig. (**6**) that there is an increase in the percentage moisture content for a specific composite as the duration of immersion in boiling water increases. For example, for 2% tiger nut fibre-epoxy composite, the percentage moisture contents are 60, 65, and 74 for immersion in boiling water for 4, 24, and 168 hr, respectively, while for 2% tiger nut fibre-nanoclay epoxy composite, the percentage moisture contents are 40, 42, and 45% for immersion in boiling water for 4, 24, and 168 hr, respectively. Also, a decreasing trend was observed in the percentage moisture contents in Fig. (**6**) as the percentage weight fractions of the reinforcement increased with respect to the duration of the immersion. For example, for 2, 4, and 6% tiger nut fibre-epoxy composites at 4 hr immersion in boiling water, the percentage moisture contents were 60, 52, and 43%, respectively, and the same was applicable to tiger nut fibre-nanoclay-epoxy composites with respect to the duration of the immersion in boiling water. The increase in the percentage moisture content for a specific composite as the duration of immersion increases is due to degradation in the cellulose cell wall of the tiger nut fibre caused by pyrolysis, leading to an increase in the inflow of water into the composite [3, 74]. Pyrolysis is a chemical process whereby the cellulose cell wall of a plant decomposes at high temperatures without oxygen [3]. Therefore, the immersion of the composites in boiling water at extended durations has the tendency to increase the rate of pyrolysis. The decrease in the percentage of the moisture content as the percentage weight fraction of the reinforcement increases was traced to an increase in the aspect ratio or ratio of length to diameter (l/d), which may have impacted the modulus of the fibres. The increase in aspect ratio reduces the rate of pyrolysis taking place and *vice versa* [3, 74]. Comparing the reduction of water absorption of tiger nut fibre-epoxy composites and tiger-nanoclay-epoxy composites as the percentage weight fractions of the reinforcement increases, the presence of nanoclay reduces the moisture absorption more by enhancing the adhesion between the fibres and the matrix [75]. The nanoclay as a compatibility agent was able to overcome the two major mechanisms attributed to water absorption in natural fibres reinforced polymer composites (NFRPCs) [76 - 79]. Table 5 shows the volume of the moisture uptake for varying weight fractions of the reinforcements.

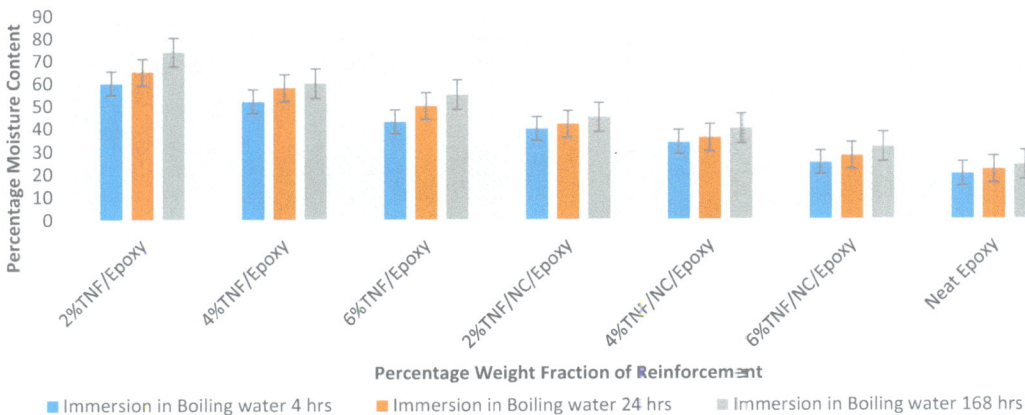

Fig. (7). Percentage moisture content against varying weight fractions of the composites.

Table 5. Percentage moisture content of the composites.

S/N	% wt. fraction	Immersion in boiling water for 4 hrs	Immersion in boiling water for 24 hrs	Immersion in boiling water for 168 hrs
1	2% TNF/Epoxy	60	65	74
2	4% TNF/Epoxy	52	58	60
3	6% TNF/Epoxy	43	50	55
4	2% TNF/NC/Epoxy	40	42	45
5	4% TNF/NC/Epoxy	34	36	40
6	6% TNF/NC/Epoxy	25	28	32
7	Neat Epoxy	20	22	24

Microstructural Analysis

Figs. (**8a** - **8f**) shows the microstructural images of tiger nut fibre-epoxy composites and tiger nut -nanoclay-epoxy composites at varying weight fractions of the reinforcements. It was observed that there were more voids in tiger nut fibre-epoxy composites leading to interfacial de-bonding, unlike in the tiger nut fibre-nanoclay-epoxy composites where the tiger nut fibre are well trapped within the polymer. The compatibility effect of the dispersed nanoclay into the matrix reduces the interfacial tension by enhancing good fibre-matrix adhesion. More so, there is an alteration in the viscosity ratio between the fibres and the polymer due to the initiation of strong interaction between the polymer chains and the solid fibres. This is as a result of the dispersion of nanoclay along with the fibre-polymer phases, which induces steric hindrance and invariably reduces the formation of voids leading to better mechanical properties of the resulting composites [3, 74, 76 - 80].

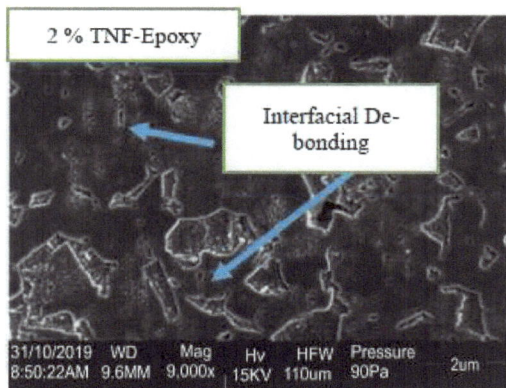

Fig. (8a). SEM micrograph of 2% TNF-epoxy composite.

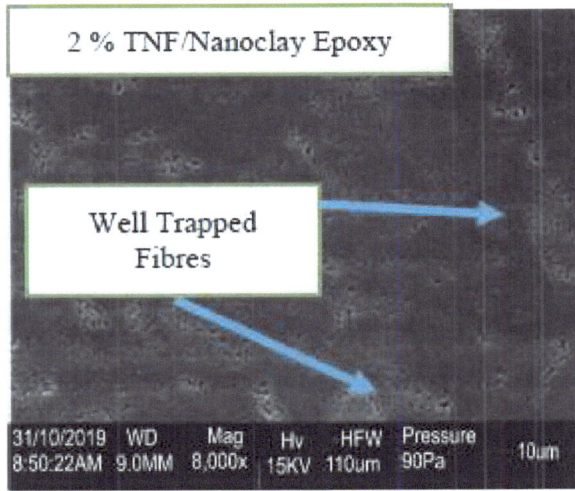

Fig. (8b). SEM micrograph of 2% TNF/Nanoclay-epoxy composite.

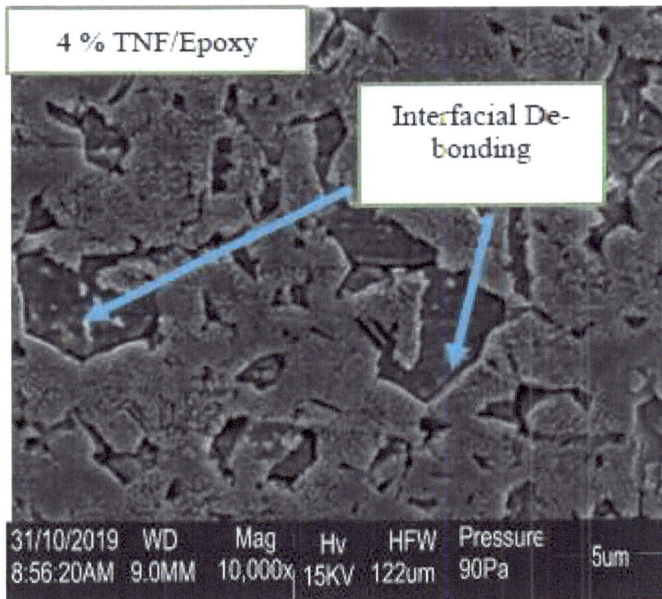

Fig. (8c). SEM micrograph of 4% TNF-epoxy composite.

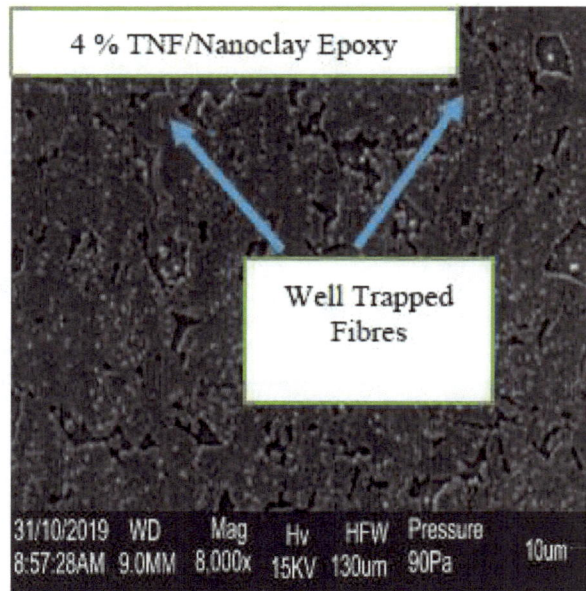

Fig. (8d). SEM micrograph of 4% TNF/nanoclay-epoxy composite.

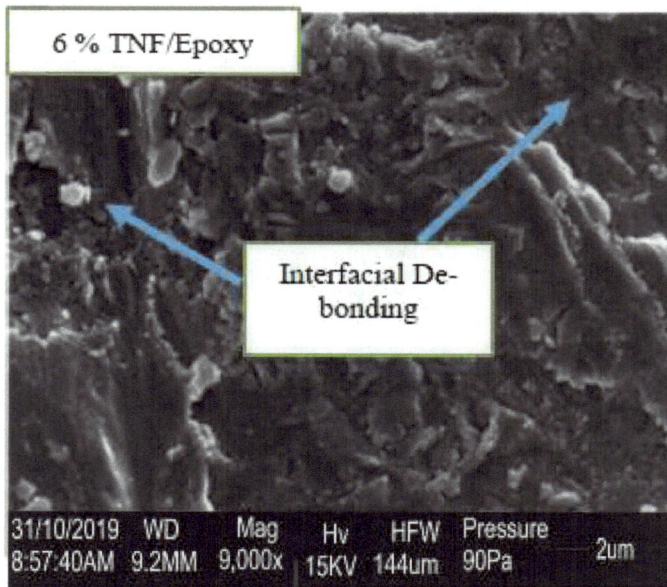

Fig. (8e). SEM micrograph of 6% TNF-epoxy composite.

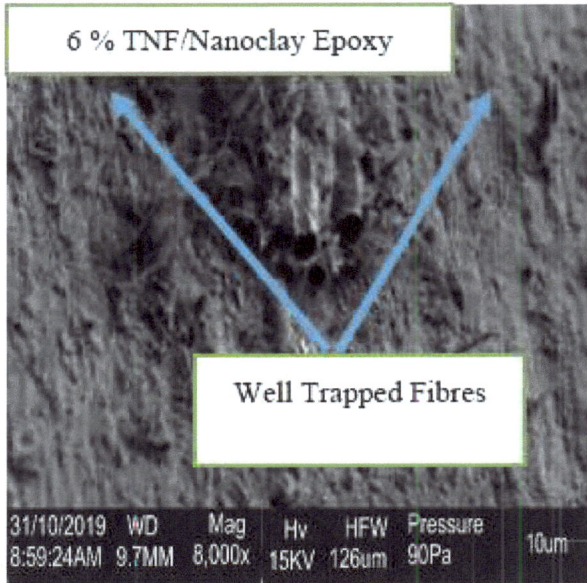

Fig. (8f). SEM micrograph of 6% TNF/nanoclay-epoxy composite.

CONCLUSION

In relation to the study objectives, the following conclusions were drawn from the results obtained.

1. Tiger nut fibre-epoxy nanocomposites were successfully developed *via* high shear mixing by adding nanoclay as a compatibility agent.
2. The effect of nanoclay mixed with tiger nut fibre for the enhancement of tensile strength, elongation, hardness, moisture uptake, and microstructure of the composites is significant.
3. The percentage enhancement in the tensile strength due to the addition of nanoclay was observed to be 31, 26, and 35% for 2, 4, and 6% weight fractions of the reinforcement, respectively.
4. The percentage enhancement in the elongation was 36, 23, and 20% for 2, 4, and 6% weight fractions of the reinforcement, respectively.
5. The percentage enhancement in the hardness were 9, 12, and 11% for 2, 4, and 6% weight fractions of the reinforcement, respectively.
6. Based on the sustainability, availability, cost effectiveness, and good mechanical properties of tiger nut fibre composites investigated, it can be considered a very promising material for the fabrication of lightweight materials used in automotive and other industrial sectors.

CONSENT FOR PUBLICATION

Not applicable.

CONFLICT OF INTEREST

The authors declare no conflict of interest, financial or otherwise.

ACKNOWLEDGEMENTS

Declared none.

REFERENCES

[1]   Y. Coskuner, R. Ercan, E. Karababa, and A.N. Nazlican, "Physical and chemical properties of chufa (Cyperus esculentus L) tubers grown in the ukurova region of Turkey", *J. Sci. Food Agric.,* vol. 82, pp. 625-631, 2002.
[http://dx.doi.org/10.1002/jsfa.1091]

[2]   R. Kumar, "M. R. UlHaq, A. Raina and A. Anand, "Industrial applications of natural fibre-reinforced polymer composites-challenges and opportunities", *Int. J. Sustain. Eng.,* vol. 12, no. 3, pp. 212-220, 2018.
[http://dx.doi.org/10.1080/19397038.2018.1538267]

[3]   C.W. Nguong, S.N.B. Lee, and D.A. Sujan, "Review on natural fibre-reinforced polymer composites", *International Journal of Material and Metallurgical Engineering,* vol. 7, no. 1, pp. 52-59, 2013.

[4]   S. Dixit, R. Goel, A. Dabey, P.R. Shivhare, and R. Bhalavi, "Natural fibre reinforced polymer composite materials- A review", *Polymer from Renewable Resources,* vol. 8, no. 2, pp. 71-78, 2017.
[http://dx.doi.org/10.1177/204124791700800203]

[5]   K.P. Ashik, and S.R. Sharma, "A review on mechanical properties of natural fibre reinforced hybrid polymer composites", *J. Miner. Mater. Charact. Eng.,* vol. 3, pp. 420-426, 2015.
[http://dx.doi.org/10.4236/jmmce.2015.35044]

[6]   S. Taj, A.M. Munawar, and K. Shafiullah, "Natural fiber-reinforced polymer composites", *Proc. Pakistan Acad. Sci.,* vol. 44, no. 2, pp. 129-144, 2007.

[7]   R AYoung, "Structure, Swelling and Bonding of Cellulose Fibers", In: *Cellulose: Structure, Modification and Hydrolysis 91-128* Wiley and Sons: New York, 1986.

[8]   S. Mohanty, S.K. Nayak, S.K. Verma, and S.S. Tripathy, "Effect of MAPP as a coupling agent on the performance of jute–PP Composites", *J. Reinf. Plast. Compos.,* vol. 23, no. 6, pp. 625-637, 2004.
[http://dx.doi.org/10.1177/0731684404032868]

[9]   N.E. Zafeiropoulos, D.R. Williams, C.A. Baillie, and F.L. Matthews, "Engineering and characterization of the interface in flax fibre/polypropylene", *Compos., Part A Appl. Sci. Manuf.,* vol. 33, no. 8, pp. 1083-1093, 2002.
[http://dx.doi.org/10.1016/S1359-835X(02)00082-9]

[10]   K.V. Sarkanen, and C.H. Ludwig, *Lignins: Occurrence, Formation, Structure and Reactions.* Wiley Interscience: New York, 1971.

[11]   W.A. Hussain, and S.N. Rafiq, "Mechanical properties of carrot fibre-epoxy composite", *Baghdad Science Journal,* vol. 9, no. 2, pp. 335-340, 2011.

[12]   A. Shalwan, and B.F. Yousif, "In state of art: Mechanical and tribological behavior of polymeric composites based on natural fibres", In: *Materials & Design* vol. 48. , 2013, pp. 14-24.

[13] J. Holbery, and D. Houston, "Natural-fiber-reinforced polymer composites in automotive applications", *JOM Journal of the Minerals, Metals and Materials Society,* vol. 58, no. 11, pp. 80-86, 2006.
[http://dx.doi.org/10.1007/s11837-006-0234-2]

[14] H.N. Dhakal, Z.Y. Zhang, and M.O.W. Richardson, "Effect of water absorption on the mechanical properties of hemp fibre reinforced unsaturated polyester composites", In: *Composites Science and Technology* vol. 67. , 2007, pp. 7-8.
[http://dx.doi.org/10.1016/j.compscitech.2006.06.019]

[15] A.K. Bledzki, and J. Gassan, "Composites reinforced with cellulose based fibres", *Prog. Polym. Sci.,* vol. 24, no. 2, pp. 221-274, 1999.
[http://dx.doi.org/10.1016/S0079-6700(98)00018-5]

[16] J. Giancaspro, C. Papakonstantinou, and P. Balaguru, "Mechanical behavior of fire-resistant biocomposite", *Compos., Part B Eng.,* vol. 40, no. 3, pp. 206-211, 2009.
[http://dx.doi.org/10.1016/j.compositesb.2008.11.008]

[17] D. Hull, and T.W. Clyne, *An Introduction to Composite Materials* Cambridge University press: Cambridge, UK, 1996.
[http://dx.doi.org/10.1017/CBO9781139170130]

[18] P.K. Mallick, *Materials, Design and Manufacturing for Lightweight Vehicles.* Woodhead Publishing Limited: Cambridge, UK, 2010.
[http://dx.doi.org/10.1533/9781845697822]

[19] D. Saravana Bavan, and G.C. Mohan Kumar, "Potential use of natural fiber composite materials in india", *J. Reinforc. Plast. Composit,* vol. 29, pp. 3600-3613, 2010.
[http://dx.doi.org/10.1177/0731684410381151]

[20] D.N. Saheb, and J.P. Jog, "Natural fiber polymer composites: A review", *Advances in Polymer Technology: Journal of the Polymer Processing Institute,* vol. 18, pp 351-363, 1999.
[http://dx.doi.org/10.1002/(SICI)1098-2329(199924)18:4<351::AID-ADV6>3.0.CO;2-X]

[21] J.K. Pandey, S.H. Ahn, C.S. Lee, A.K. Mohanty, and M. Misra, "Recent Advances in the Application of Natural Fiber Based Composites", *Macromol. Mater. Eng.,* vol. 295, pp. 975-989, 2010.
[http://dx.doi.org/10.1002/mame.201000095]

[22] S. Das, "Jute Composite and Its Applications", *Proceedings of International Workshop IJSG.* Dhaka, Bangladesh

[23] R.K. Nakamura, J.N. Goda, and X.J. Ohgi, "High Temperature Tensile Properties and Deep Drawing of Fully Green Composites", *Express Polym. Lett.,* vol. 3, p. 1, 2009.
[http://dx.doi.org/10.3144/expresspolymlett.2009.4]

[24] A. Alegria-Toran, R. Farre-Rovira, and Y.S. Horchata, "Aspectos nutricionales y diet´eticos. In: Fundaci´on Valenciana de Estudios Avanzados", editor. Jornada Chufa y Horchata: Tradici´on y Salud", In: *Valencia, Spain: Conseller´ia de Agricultura, Pesca y Alimentaci*, 2003, pp. 55-70.

[25] R.M. Rowell, "Property Enhanced Natural Fiber Composite Materials Based on Chemical Modification", In: *Science and Technology of Polymers and Advanced Materials.,* P.N. Prasad, Ed., Plenum Press: New York, 1998, pp. 717-732.
[http://dx.doi.org/10.1007/978-1-4899-0112-5_63]

[26] E. Zini, and M. Scandola, "Green composites: An overview", *Polym. Compos.,* vol. 32, no. 12, pp. 1905-1915, 2011.
[http://dx.doi.org/10.1002/pc.21224]

[27] H. Alamri, and I.M. Low, "Effect of water absorption on the mechanical properties of n-sic filled recycled cellulose fiber reinforced epoxy eco-nanocomposites", *Polym. Test.,* vol. 06, no. 001, pp. 810-818, 2012.
[http://dx.doi.org/10.1016/j.polymertesting.2012.06.001]

[28] H. Alamri, I.M. Low, and Z. Alothman, "Mechanical, Thermal and Microstructural Characteristics of Cellulose Fiber Reinforced Epoxy/OrganoclayNanocomposites", *Compos., Part B Eng.,* vol. 04, no. 037, pp. 2762-2771, 2012.
[http://dx.doi.org/10.1016/j.compositesb.2012.04.037]

[29] D. Ray, and J. Rout, *Thermoset Biocomposites.,* A. K. Mohanty, M. Misra , L.T. Drzal, Eds., New York: Taylor & Francis Group, 2005.
[http://dx.doi.org/10.1201/9780203508206.ch9]

[30] M.S. Nazir, and M.H.M. Kassim, "M.H.M, L. Mohapatra, M.A. Gilani, M.R. Raza, K. Majeed", In: *Characteristic Properties of Nanoclays and Characterization of Nanoparticulates and Nanocomposites,* R.P.C. Nanoclay, Ed., Springer: Singapore, 2016, pp. 35-55.

[31] K. Müller, E. Bugnicourt, M. Latorre, M. Jorda, Y. Echegoyen Sanz, J.M. Lagaron, O. Miesbauer, A. Bianchin, S. Hankin, U. Bölz, G. Pérez, M. Jesdinszki, M. Lindner, Z. Scheuerer, S. Castelló, and M. Schmid, "Review on the processing and properties of polymer nanocomposites and nanocoatings and their applications in the packaging, automotive and solar energy fields", *Nanomaterials (Basel),* vol. 7, no. 4, p. 74, 2017.
[http://dx.doi.org/10.3390/nano7040074] [PMID: 28362331]

[32] G.J.M. Rytwo, "Clay minerals as an ancient nanotechnology: historical uses of clay organic interactions, and future possible perspectives", *Macla,* vol. 9, pp. 15-17, 2008.

[33] S.M. Lee, and D. Tiwari, "Organo and inorgano-organo-modified clays in the remediation of aqueous solutions: An overview", *Appl. Clay Sci.,* vol. 59, pp. 84-102, 2012.
[http://dx.doi.org/10.1016/j.clay.2012.02.006]

[34] M.K. Uddin, "A review on the adsorption of heavy metals by clay minerals, with special focus on the past decade", *Chem. Eng. J.,* vol. 308, pp. 438-462, 2017.
[http://dx.doi.org/10.1016/j.cej.2016.09.029]

[35] M. Jawaid, A.K. Qaiss, and R. Bouhfid, *Nanoclay Reinforced Polymer Composites: Nanocomposites and Bionanocomposites.* Springer: Singapore, 2016.
[http://dx.doi.org/10.1007/978-981-10-1953-1]

[36] K. Majeed, M. Jawaid, A. Hassan, A. Abu Bakar, H.P.S. Abdul Khalil, A.A. Salema, and I. Inuwa, "Potential materials for food packaging from nanoclay/natural fibres filled hybrid composites", *Mater. Des.,* vol. 46, pp. 391-410, 2013.
[http://dx.doi.org/10.1016/j.matdes.2012.10.044]

[37] E.P. Giannelis, "Polymer layered silicate nanocomposites", *Adv. Mater.,* vol. 8, pp. 29-35, 1996.
[http://dx.doi.org/10.1002/adma.19960080104]

[38] K. Yusoh, S.V. Kumaran, and F.S. Ismail, "Surface Modification of Nanoclay for the Synthesis of Polycaprolactone (PCL)—Clay Nanocomposite", *Proceedings of the MATEC Web of Conferences,* 2017 Penang, Malaysia

[39] M.R. Irshidat, and M.H. Al-Saleh, "Thermal performance and fire resistance of Nanoclay modified cementitious materials", *Constr. Build. Mater.,* vol. 159, pp. 213-219, 2018.
[http://dx.doi.org/10.1016/j.conbuildmat.2017.10.127]

[40] G. Choudalakis, and A. Gotsis, "Permeability of polymer/clay nanocomposites: A review", *Eur. Polym. J.,* vol. 45, pp. 967-984, 2009.
[http://dx.doi.org/10.1016/j.eurpolymj.2009.01.027]

[41] S. Ganguly, K. Dana, T.K. Mukhopadhyay, T. Parya, and S. Ghatak, "Organophilic nano clay: A Comprehensive review", *Trans. Indian Ceram. Soc.,* vol. 70, pp. 189-206, 2011.
[http://dx.doi.org/10.1080/0371750X.2011.10600169]

[42] N. Öztürk, A. Tabak, S. Akgöl, and A. Denizli, "Newly synthesized bentonite–histidine (Bent–His) micro-composite affinity sorbents for IgG adsorption", *Colloids Surf. A Physicochem. Eng. Asp.,* vol. 301, pp. 490-497, 2007.

[http://dx.doi.org/10.1016/j.colsurfa.2007.01.026]

[43]   S. Pavlidou, and C. Papaspyrides, "A review on polymer–layered silicate nanocomposites", *Prog. Polym. Sci.,* vol. 33, pp. 1119-1198, 2008.
[http://dx.doi.org/10.1016/j.progpolymsci.2008.07.008]

[44]   P. Liu, "Polymer modified clay minerals: A review", *Appl. Clay Sci.,* vol. 38, pp. 64-76, 2007.
[http://dx.doi.org/10.1016/j.clay.2007.01.004]

[45]   A.H. Ambre, K.S. Katti, and D.R. Katti, "Nanoclay based composite scaffolds for bone tissue engineering applications", *J. Nanotechnol. Eng. Med.,* vol. 1, no. 031013, 2010.
[http://dx.doi.org/10.1115/1.4002149]

[46]   M.I. Carretero, and M. Pozo, "Clay and non-clay minerals in the pharmaceutical and cosmetic industries Part II. Active ingredients", *Appl. Clay Sci.,* vol. 47, pp. 171-181, 2010.
[http://dx.doi.org/10.1016/j.clay.2009.10.016]

[47]   S. Shahidi, and M. Ghoranneviss, "Effect of Plasma Pretreatment Followed by Nanoclay Loading on Flame Retardant Properties of Cotton Fabric", *J. Fusion Energy,* vol. 33, pp. 88-95, 2014.
[http://dx.doi.org/10.1007/s10894-013-9645-6]

[48]   S.S. Ray, and M. Okamoto, *Polymer/layered Silicate Nanocomposites: A Review from Preparation to Processing* vol. 28. Prog. Polym. Sci., 2003, pp. 539-1641.

[49]   W. Gacitua, A. Ballerini, and J. Zhang, "Polymer nanocomposites: Synthetic and natural fillers a review", *Maderas Cienc. Tecnol.,* vol. 7, pp. 159-178, 2005.
[http://dx.doi.org/10.4067/S0718-221X2005000300002]

[50]   V.S. Vo, S. Mahouche-Chergui, V.H. Nguyen, S. Naili, N.K. Singha, and B. Carbonnier, "Chemical and Photochemical Routes toward Tailor-Made Polymer–Clay Nanocomposites: Recent Progress and Future Prospects", In: *Clay-Polymer Nanocomposites,* K. Jlassi, M.M. Chehimi, S. Thomas, Eds., Elsevier: Amsterdam, the Netherlands, 2017, pp. 145-197.
[http://dx.doi.org/10.1016/B978-0-323-46153-5.00005-7]

[51]   G. Caprino, L. Carrino, M. Durante, A. Langella, and V. Lopresto, ' Low impact behaviour of hemp fibre reinforced epoxy composites", In: *Composite Structures* vol. 133. , 2015, pp. 892-901.

[52]   K.G. Satyanarayana, G.G.C. Arizaga, and F. Wypych, "Biodegradable composites based on lignocellulosic fibers an overview", *Prog. Polym. Sci.,* vol. 34, no. 9, pp. 982-1021, 2009.
[http://dx.doi.org/10.1016/j.progpolymsci.2008.12.002]

[53]   A.K. Mohanty, M. Misra, and H.G. Biofibres, "biodegradable polymers and biocomposites: An overview", *Macromol. Mater. Eng.,* vol. 276, no. 1, pp. 1-24, 2000.
[http://dx.doi.org/10.1002/(SICI)1439-2054(20000301)276:1<1::AID-MAME1>3.0.CO;2-W]

[54]   A. Anand, M. Irfan Ul Haq, K. Vohra, A. Raina, and M.F. Wani, "Role of green tribology in sustainability of mechanical systems: A state of the art survey", *Mater. Today Proc.,* vol. 4, no. 2, pp. 3659-3665, 2017.
[http://dx.doi.org/10.1016/j.matpr.2017.02.259]

[55]   M. Karus, and M. Kaup, "Natural fibres in the European automotive industry", *J. Ind. Hemp,* vol. 7, no. 1, pp. 119-131, 2002.
[http://dx.doi.org/10.1300/J237v07n01_10]

[56]   R.V. Silva, D. Spinelli, W.W. Bose Filho, S. Claro Neto, G.O. Chierice, and J.R. Tarpani, "Fracture toughness of natural fibers/castor oil polyurethane composites", *Compos. Sci. Technol.,* vol. 66, no. 10, pp. 1328-1335, 2006.
[http://dx.doi.org/10.1016/j.compscitech.2005.10.012]

[57]   T.G. Schuh, *Renewable Materials for Automotive Applications* Daimler-Chrysler AG: Stuttgart, 1999.

[58]   H.A. Al-Qureshi, "The Use of Banana Fibre Reinforced Composites for the Development of a Truck Body", *Second International Wood and Natural Fibre Composites Symposium,* 1999, pp. 1-8

Kassel/Germany.

[59]   M.M. Davoodi, S.M. Sapuan, D. Ahmad, A. Ali, A. Khalina, and M. Jonoobi, "Mechanical properties of hybrid kenaf/glass reinforced epoxy composite for passenger car bumper beam", In: *Materials & Design*, 2010.
[http://dx.doi.org/10.1016/j.matdes.2010.05.021]

[60]   C.A. Alves, J. Silva, L.G. Reis, M. Freitas, L.B. Rodrigues, and D.E. Alves, "Ecodesign of automotive components making use of natural jute fiber composites", *J. Clean. Prod.*, vol. 18, no. 4, pp. 313-327, 2010.
[http://dx.doi.org/10.1016/j.jclepro.2009.10.022]

[61]   M. Saxena, M.J. Nandan, and N. Ramakrishan, "Sisal: Potential for employment generation and rural development", *IOSR J. Mech. Civ. Eng.*, vol. 8, no. 6, pp. 1-8, 2011.

[62]   H.Q. Pham, and M.J. Marks, Epoxy Resins.*Encyclopedia of Polymer Science and Technology,* H.F. Mark, Ed., 3rd ed. John Wiley & Sons, 2004.
[http://dx.doi.org/10.1002/0471440264.pst119]

[63]   M.E. Hossain, M.K. Hossain, M.V. Hosur, and S. Jeelani, "Study of mechanical responses and thermal expansion of cnf-modified polyester nanocomposites processed by different mixing systems", *Mater. Res. Soc. Symp. Proc. Proceedings*, vol. 1312, . pp. 225-230.
[http://dx.doi.org/10.1557/opl.2011.111]

[64]   M.K. Hossain, M.E. Hossain, M.V. Hosur, and S. Jeelani, "Flexural and compression response of woven e-glass/polyester–cnf nanophased composites", *Comp: Part A.*, vol. 42, no. 11, pp. 1774-1782, 2011.
[http://dx.doi.org/10.1016/j.compositesa.2011.07.033]

[65]   P.S. Mukherjee, and K.G. Satyanarayana, "Structure and properties of some vegetable fibres", *J. Mater. Sci*, vol. 19, pp. 3925-3934, 1984.
[http://dx.doi.org/10.1007/BF00980755]

[66]   N. Reddy, and Y. Yang, "Biofibers from agricultural byproducts for industrial applications", *Trends Biotechnol.*, vol. 23, no. 1, pp. 22-27, 2005.
[http://dx.doi.org/10.1016/j.tibtech.2004.11.002] [PMID: 15629854]

[67]   H.S. Yang, H.J. Kim, B.J. Lee, and T.S. Hawng, "Rice husk flour filled polypropylene composites; mechanical and morphological study", *Compos. Struct.*, vol. 63, pp. 305-312, 2004.
[http://dx.doi.org/10.1016/S0263-8223(03)00179-X]

[68]   H.S. Yang, H.J. Kim, B.J. Lee, and T.S. Hawng, "Effect of compatibilizing agent on rice husk flour reinforced polypropylene composites", *Compos. Struct.*, vol. 77, pp. 45-55, 2007.
[http://dx.doi.org/10.1016/j.compstruct.2005.06.005]

[69]   T.P. Mohan, and K. Kanny, "Water Barrier Properties of Nanoclay Filled Sisal Reinforced Epoxy Composites", *Compos., Part A Appl. Sci. Manuf.*, vol. 12, no. 010, pp. 385-393, 2010.

[70]   C.V. Srinivasa, and K.N. Bharath, "Impact and hardness properties of areca fibre-epoxy reinforced composites", *Journal of Material Environmental Science*, vol. 2, no. 4, pp. 351-356, 2011.

[71]   E.C. Ramires, and E. Frollini, "Tannin-phenolic resins: synthesis, characterization and application in bio based composites reinforced with sisal fibres", *Compos., Part B Eng.*, vol. 43, no. 7, pp. 2836-2842, 2012.
[http://dx.doi.org/10.1016/j.compositesb.2012.04.049]

[72]   W. Webo, L. Masu, and M. Maringa, "The impact toughness and hardness of treated and untreated sisal fibre-epoxy resin composites", In: *Advances in Materials Science and Engineering*, 2018, pp. 1-10.
[http://dx.doi.org/10.1155/2018/8234106]

[73]   Y. Han-Seung, M.P. Walcott, S.K. Hee-sookim, and K. Hyunjoong, "Effect of different compatibilizing agents on the mechanical properties of lignocellulosic material filled polyethylene bio

composites", *Journal of Composite Structures.,* vol. 79, no. 3, pp. 369-375, 2007.
[http://dx.doi.org/10.1016/j.compstruct.2006.02.016]

[74]   K.S. Meenalochani, and B.G. Reddy, "A review of water absorption behaviour and its effect on mechanical properties of natural fibres reinforced composites", *International Journal of Innovative Research in Advanced Engineering,* vol. 4, no. 4, pp. 143-147, 2017.

[75]   *Halip, L.S. Hua, Z. Ashaari, P.M. Tahir, L.W. Chen and M.K.A. Uyup. "Effect of Treatment on Water Absorption Behaviour of Natural Fibre-Reinforced Polymer Composites". Mechanical and Physical Testing of Biocomposites, Fibre-Reinforced Composites and Hybrid Composites* Elsevier Ltd, 2019.

[76]   A.R. Kakroodi, Y. Kazemi, and D. Rodrigue, "Mechanical, rheological, morphological and water absorption properties of maleated polyethylene/hemp composites: effect of ground tire rubber addition", *Compos Part B,* vol. 51, p. 337e44, 2013.

[77]   B.K. Kim, O.H. Kwon, W.H. Park, and D. Cho, *Thermal, mechanical, impact, and water absorption properties of novel silk fibroin fiber reinforced poly (butylene succinate) biocomposites* vol. 24. Macromol Res, 2016, pp. 734-740.
[http://dx.doi.org/10.1007/s13233-016-4102-9]

[78]   G. Kalaprasad, B. Francis, S. Thomas, C.R. Kumar, C. Pavithran, G. Groeninckx, and S. Thomas, "Effect of fibre length and chemical modifications on the tensile properties of intimately mixed short sisal/glass hybrid fibre reinforced low density polyethylene composites", In: *Polymer Int* vol. 53. , 2004, no. 11, pp. 1624-1638.
[http://dx.doi.org/10.1002/pi.1453]

[79]   M.N. Ichazo, C. Albano, J. Gonzalez, R. Perera, and M.V. Candal, "Polypropylene/wood flour composites: treatments and properties", In: *Compos Struct,* 2001.

[80]   J. Karami, H. Nazockdast, Z. Ahmadi, J.F. Rabolt, I. Noda, and D.B. Chase, "Microstructure effects on the rheology of nanoclay-filled PHB/LDPE blends", *Polym. Compos.,* pp. 1-10, 2019.
[http://dx.doi.org/10.1002/pc.25273]

Advances in Manufacturing Technologies, 2022, 45-55

Assessment of Microstructure and Mechanical Properties of As-cast Magnesium Alloys Reinforced with Organically Extracted Zinc and Calcium

Temitayo Mufutau Azeez[1,2,*], **Lateef Owolabi Mudashiru**[2] and **Abiodun Ayodeji Ojetoye**[3]

[1] *Department of Mechanical and Mechatronic Engineering, Afe Babalola University, Ado-Ekiti, Nigeria*

[2] *Department of Mechanical Engineering, Ladoke Akintola University of Technology, Ogbomoso, Nigeria*

[3] *Department of Mechanical Engineering, University of Ibadan, Ibadan, Nigeria*

Abstract: Magnesium (Mg) is commonly used as a biomaterial because of its biocompatibility, biodegradation, non-toxicity, and good mechanical properties. The conventional consideration for selecting Mg alloy elements is based on their corrosion resistance, good hardness, and strength. Therefore, some of the alloying elements that enable the above properties are zinc, calcium, and aluminum. The future requirement of biomaterial involves non-toxicity in addition to the existing properties. The tensile strength and hardness of polymeric materials used as a replacement for metallic materials in biomedical applications were lesser. Therefore, calcium and zinc were sourced organically from cow bones and cocoa seeds, respectively, to eliminate the adverse effect of the inorganic source of the same element. Zinc and calcium were alloyed with magnesium at different percentages recommended by Junjin, 2017 experiments (0.23% Zn, 0.15% Ca in alloy 1 and 0.25% Zn, 0.23% Ca in alloy 2). The hardness, tensile strength, and percentage elongation of the as-cast materials from organic alloy sources were investigated and compared to similar experiments from the literature with the inorganic alloy source. The result showed that Mg alloy with organic zinc and calcium exhibit better hardness (52.1 HRv in alloy 1 and 60.7 HRv in alloy 2), tensile strength (181.3 MPa in alloy 1 and 208.3MPa in alloy 2), and ductility (13.1% elongation in alloy 1 and 18% in alloy 2) compared to Jinjin Mg alloy from an inorganic source of lesser values of mechanical properties. It can be concluded that zinc and calcium from organic sources is a better replacement for inorganic sources in Mg alloy.

Keywords: Alloying elements, Biomaterial, Magnesium, Mechanical properties.

* **Corresponding Author Temitayo Mufutau Azeez:**Department of Mechanical and Mechatronic Engineering, Afe Babalola University, Ado-Ekiti, Nigeria and Department of Mechanical Engineering, Ladoke Akintola University of Technology, Ogbomoso, Nigeria; Tel: +2348060800687; E-mail: lomudashiru@lautech.edu.ng

Ilesanmi Afolabi Daniyan (Ed.)

INTRODUCTION

Pure magnesium (Mg), other elements such as Zn, Mn, Al, Ca, Li, Zr, Y, and rare earth metals (RE) have been used [1]. Magnesium has emerged as a potential biomaterial owing to its biocompatibility, good mechanical strength, and biodegradation. Mg is nontoxic [2], with the daily recommended intake of 240-420 mg/day for adults [3]. This value is almost 50 times higher than the recommended intake of iron (Fe) and Zinc (Zn), which are other potential implant materials [4]. Additionally, Mg and its alloys have shown excellent biocompatibility in physiological conditions [5]. Along with biocompatibility, Mg has suitable mechanical properties for an implant material due to its lightweight and good strength to weight ratio [2]. Moreover, Mg's elastic modulus is about 45 GPa, which is closer to the elastic modulus of bone (3-20 GPa), reducing the possibility of stress shielding [6].

On the other hand, Fe and Zn have elastic modulus values of 211.4 and 90 GPa, respectively, much higher than bone. Along with suitable mechanical properties and biocompatibility, biodegradation is the primary reason for Mg's enhanced interest as an implant material [7]. Prolonged interactions of implants in the biological surroundings can lead to many complexities that are not desirable. Mg alloy implants avoid these long-term incompatible interactions with the body tissues eliminating the possibilities of any complexity. All the above-mentioned properties make Mg a potential material to replace the conventional permanent implant materials.

Several magnesium compositions are currently being explored. Some of the alloys are more developed at research stages than others, and each alloy has been tailored for specific applications, as previously mentioned. The use of pure magnesium and other elements such as Zn, Mn, Al, Ca, Li, Zr, Y, and rare earth metals (RE) has been reported [1]. Harnessing these elements with the Mg matrix creates different mechanical and physical properties in the resulting alloy. If the alloying element can accomplish the metallurgical principle of forming a solid solution, then solid solution strengthening can be achieved. A solid-solution is formed when two or more alloys are completely soluble in one another. The formation of inter-matrix phases improves the strength of the alloy, and it is referred to as dispersion strengthening. This is a common practice when forming an alloy. Typically, one metal will have larger atoms relative to other constituents within the material and form a ductile phase called the matrix [8]. The matrix consists of the major volume for the alloy. The second metal added to the alloy consists of smaller atoms that are usually both stronger and harder. When the two metals are mixed in the formation process, the resulting material is dispersion strengthened [2]. Between the microstructure of two or more phases grain boundaries exist. When dispersion strengthening occurs, micro-scaled precipitates usually form within the grain boundaries, further strengthening the material as the precipitates prevent slippage of the dislocations or defects within the grain or phase. Such a treatment can be induced by heat treatment within the manufacturing process, and it is referred to as precipitation hardening [6].

Aluminum (Al) is a commonly used alloying element in Mg, and its concentration ranges between 2 and 9 wt. % in commercially used Mg-Al alloys [4]. Mg is alloyed with Al to increase both the strength and corrosion resistance. Alloying of Mg with Al promotes both the solid solution strengthening and precipitation strengthening, thus increasing Mg-Al alloys' strength. Along with improved strength, Mg-Al alloys also show very good castability. The reason for the improved castability of Mg-Al alloys is the lowering of solidus and liquidus lines as the percentage of Al increases. The content of Al in Mg directly affects the corrosion rate of Mg-Al alloys, and increasing the Al content typically results in lower corrosion rates.

Biomaterials are nonviable materials used in medical devices intended to interact with the biological system. A designed biomaterial should serve its purpose in the living body's environment without affecting other bodily organs [9]. For this reason, a biomaterial should be nontoxic. Toxicity for biomaterials deals with the substances that migrate out of the biomaterials. In general, non-toxicity refers to non-inflammatory and non-allergenic, among others. It is reasonable to say that a biomaterial should not give off anything from its mass unless it is specifically engineered to do so.

In some cases, the biomaterials tend to release the necessary masses considered toxic [10]. Hence the need to replace some alloying organic alloying elements with inorganic alloying elements is essential.

Conventional medical devices require the use of materials that must have been experimentally proven to be corrosion resistant when placed within certain fluids. With a necessity for next-generation biomaterials for medical applications, non-corrosive materials are being considered for applications where temporary medical device placement is desired [11]. Such cases are commonly found in orthopaedic and cardiovascular applications where temporary scaffolds for structural support are required, ultimately enhancing the wound healing and regenerative tissue process [5]. The major line of biodegradable alloys for such purposes are those based upon a magnesium-zinc and calcium platform. These alloys have been composed in various assortments and have been tailored for biomedical applications [2]. The selection of elemental compositions for the magnesium alloys should meet the requirements of material science principles for coherent material formation, but more importantly, they should yield non-corrosive by-products that have minimal caustic effects on bodily function during the degradation process. Materials composed of minerals and trace elements already existing within the body are highly advantageous for such purposes. Polymeric materials are highly acclaimed for such biomedical applications due to their biocompatibility, degradation abilities, and elastic compliance for cardiovascular applications. However, several mechanical and physical properties such as loading failure in load-bearing devices and post-implantation recoiling for cardiovascular applications are less when compared to that of metallic magnesium alloys [3]. To reduce the cost of production of biomaterial and reduce toxicity, an organic source for the selection of some trace alloying elements such as calcium and zinc needs to be considered.

MATERIALS AND METHOD

Materials Collection

Samples of dried cocoa beans were obtained from cocoa farmers at Ado Ekiti, Ekiti State, Nigeria. Some of the cocoa beans were obtained in the form of their fruits (pods). The beans from these fruits were subjected to fermentation and sun-dried until they were fully dried. Samples were then kept in clean and dried polyethylene bags, as shown in Fig. (**1a**). At the same time, samples of cow bone wastes were obtained from a slaughterhouse in Ado Ekiti, Ekiti State, Nigeria. These bones were processed to obtain calcium. Aluminum ingots were bought from an industrial mineral supplier in Ikeja, Lagos state. Magnesium samples were obtained from a distributor of solid minerals situated at Ebute Metta, Lagos State.

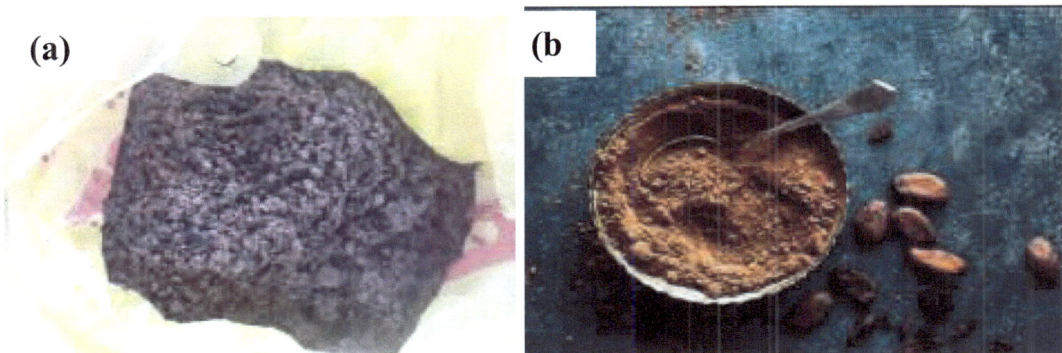

Fig. (1). Cocoa seed (**a**) raw and undried (**b**) dried and milled.

Digestion of samples and alkaline treatment for calcium extraction was performed using aqua reagents (HNO_3 and HCl) and ($NaOH$ and HCl), respectively. Deionized water was obtained from the Laboratory of Industrial Chemistry, Afe Babalola University, Ado Ekiti, Nigeria.

Zinc Sample Preparation

The dried beans were milled into fine cocoa powder, as shown in Fig. (**1b**), using a hammer mill and subsequently subjected to sample digestion. The aqua reagents for the digestion were prepared by mixing 3:1 volume of HCl and HNO_3, respectively, in a fume hood. The prepared aqua reagent was stored for two days to ensure a complete reaction and a uniform homogenous mixture between the acids before digestion of the samples commenced. 1 g of the milled cocoa sample was digested with 30 ml of the aqua reagents in a pre-cleaned Teflon cup by heating on a hot plate at 200 °C for about 20 min. The digest after cooling was transferred into a 50 ml volumetric flask.

Calcium Sample Preparation

Calcium was extracted from cow bone through the alkaline treatment method. 500 g of cow bone was dried in a hot air oven at 65 °C for 10 hr (Fig. **2**). 3% of NaOH was added to the dried cow bone. The mixture was boiled at 100 °C for 1 hr. The treated bone was separated with a filter cloth. The bone was washed twice with a solution of 1% HCl and deionized water until the neutrality was determined with pH paper. The neutralized bone was dried in a hot air oven at 100 °C for 2 hr. The dried bones were ground in a hammer mill until they passed a sieve of 100 mesh sizes.

Fig. (2). Cow bone-in hot air oven.

Melting and Casting Process of the Alloys

Aluminum alloys were prepared from pure aluminum ingots, extracted zinc, and calcium chips, and magnesium alloy. Melting, then, occured when the materials were passed through an electrical resistance furnace in the graphite crucible at 720 °C. The alloying elements were not added at once but sequentially; first, magnesium was subsequently added, and the extracted calcium element was added; zinc was also added to the alloy and mixed immediately (the zinc was not added at the same time with calcium and magnesium because of its less melting point than aluminum). The stirring of the melt was done for 5 min. and then preserved for 20 min. at the melting condition for stabilization. At last, the melting was poured into a mould, which has been prepared. The furnace used in casting processes is presented in Fig. (**3a**), the mould and the pattern used are presented in Fig. (**3b**), and the As- cast mg alloy is presented in Fig. (**3c**).

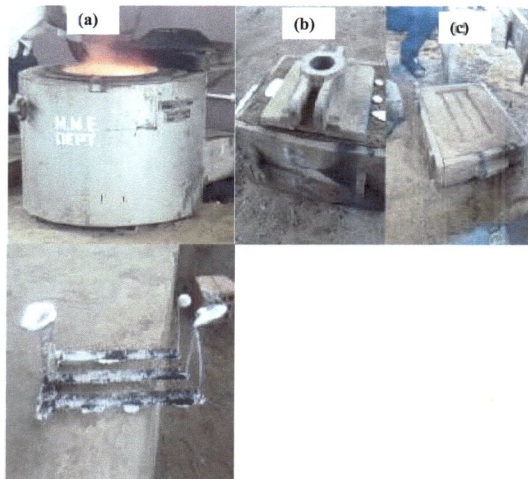

Fig. (3). Casting process (**a**) furnace used (**b**) mould and pattern used and (**c**) As-cast mg alloy.

Hardness Test

A hardness test was carried out to evaluate the as-cast and heat-treated alloys' mechanical properties. The hardness and tensile tests were performed using an Instron-5569 universal testing machine at the Department of Mechanical and Mechatronics, Afe Babalola University. Moreover, the Vickers' hardness of the prepared samples was measured with a Clark microhardness tester using a 500 g load cell and 15 sec dwell time.

Tensile Test

Tensile tests were performed on the cast samples. The cast samples were machined to 3 mm diameter, 10 mm gauge length, and 12 mm full length dimension as presented in Fig. (**4a**) for the tensile test. The tensile test was a standard test conducted on the Universal Testing Machine (UTM) following standard procedures. The sample was fixed in the tensile testing machine, and a pulling force was applied to the aluminum axially. The schematic representation of the experimental setup is presented in Fig. (**4b**). Each test was performed thrice to ensure repeatability of the result. The engineering stress-strain curves and the tensile test data were used to evaluate some tensile properties such as ultimate tensile strength (UTS), the yield strength, elastic limit, ultimate load, breaking load, percentage elongation, and percentage reduction area.

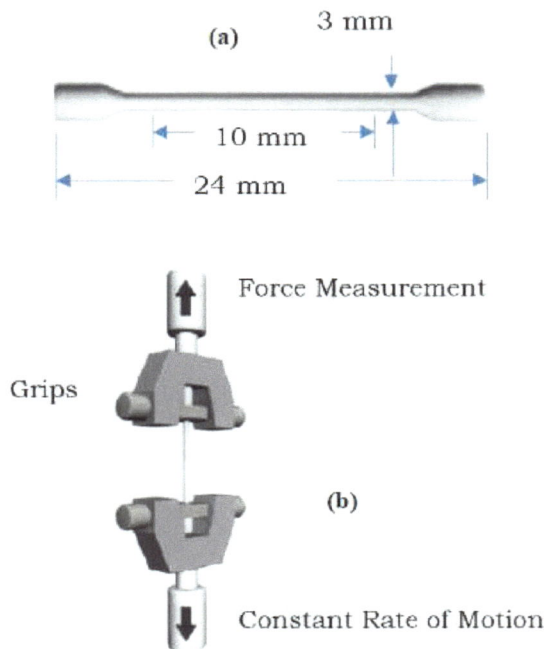

Fig. (4). Tensile test **(a)** samples dimension **(b)** schematic representation of experimental setup.

Scanning Electron Microscope (SEM)

Immediately after machining the samples, examination of the grain morphology samples started by filing these samples with a coarse file and then with a smooth file to provide the initial flatness followed by surface grinding. Preliminary machine grinding of the extruded specimen surface was followed by grinding with 1000 B grain size emery paper to achieve smooth and shinning surface using metal series model 2000, Germany. Finally, polishing with the diamond paste of grade B (3 mm) was performed until a smooth and mirror-finished surface was obtained using a polishing machine. The specimen was etched in 2% HNO_3 (Nital) for about 10 sec, dried with a methylated spirit to remove moisture, and examined on the SEM.

RESULTS AND DISCUSSIONS

Chemical Compositions of the Alloys

Table 1 shows the chemical compositions of Al-Zn-Mg-Ca alloys. The percentage variation in composition was also indicated to determine the effect of impurities in each sample cast. The percentage was varied based on the work of Junjin [1]. This helps determine the degree of closeness between an organic and inorganic source of alloying elements in its hardness and tensile strength.

Table 1. Chemical composition of the Mg-based alloy.

Alloys	Al (%)	Zn (%)	Ca (%)	Mg (%)
Alloy 1	4.03	0.23	0.15	95.59
Alloy 2	4.03	0.25	0.23	95.49

Mechanical Properties of the Al-Zn-Mg-Ca Alloys

There were significant differences in the Al-Zn-Mg-Ca alloys' mechanical properties due to the variation of organic source Zn and Ca contents in Mg alloy. The value of the tensile strength increased notably (from 177.6 to 188.3 MPa in alloy 1 and 205 MPa to 208 MPa in alloy 2). The UTS and elongation also increased with the extracted Zn content's increment. This result has shown that the organically extracted Zn and Ca improve the ductility of Mg alloy.

From the experiment, the hardness value of alloy 1 was 52.1 HRv, and the hardness value of alloy 2 was 60.7 HRv. The hardness value from the work of Junjin [1] for Mg cast alloy 1 was 49.3 HRv, and the hardness value of cast alloy 2 was 53.4 HRv, as presented in Table 2. This result showed that organically extracted Zn relatively enhanced the strength of the alloy greatly.

Table 2. Mechanical properties' comparison of experimental findings with Junjin [1].

Properties	Junjin Cast Mg alloy		As-cast mg Alloy	
-	Alloy 1	Alloy 2	Alloy 1	Alloy 2
Hardness (Hv)	49.30	53.40	52.10	60.70
Tensile Strength (MPa)	177.60	205.00	181.3	208.30
Elongation Percentage	10.80	14.00	13.10	18.00

Similarly, the tensile strength value of alloy 1 was 181.3 MPa with 13.10% elongation, and the tensile strength of alloy 2 was 208.3 MPa with 18%

elongation (Table **2**). The tensile strength value of the cast Mg alloy reported by Junjin [1] was 177.6 with 10.8% elongation, while that of alloy 2 was 205 MPa with 14.7% elongation. This result showed that organically extracted Zn relatively enhances the strength of the Mg alloy greatly. The comparison of the experimental findings from this study and that of Junjin [1] in terms of mechanical properties are further summarized in Fig. (**5**).

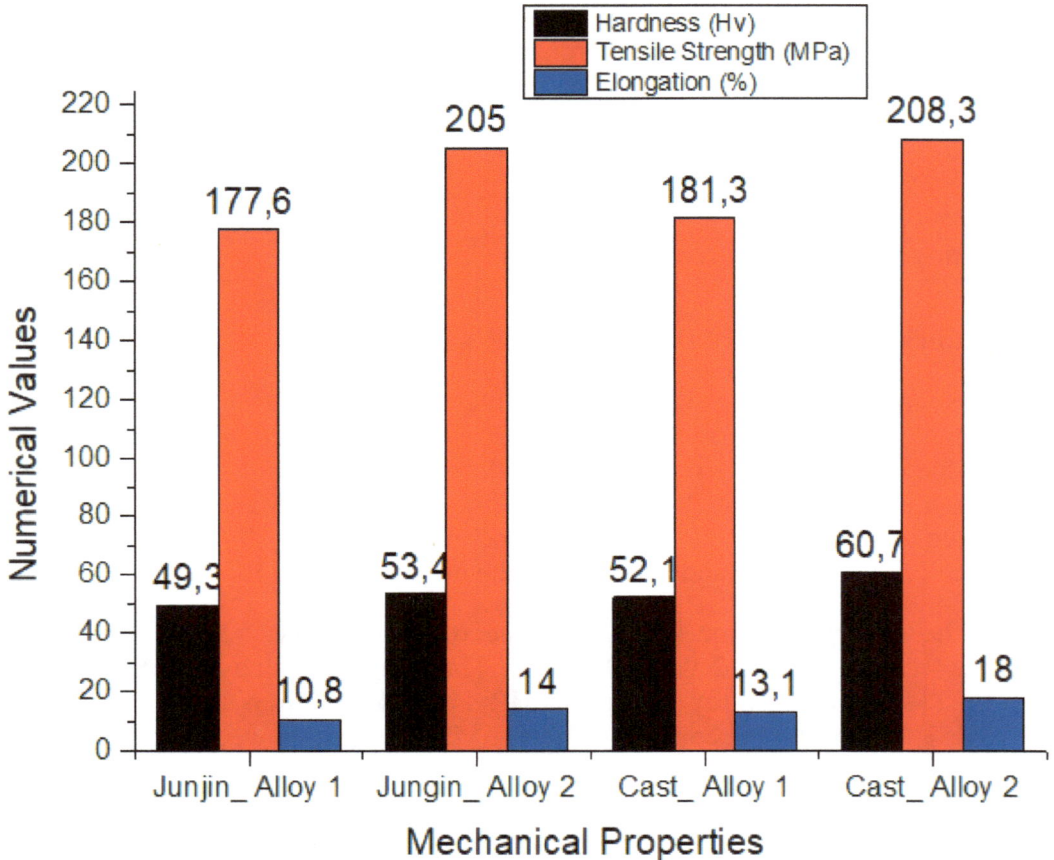

Fig. (5). Tensile strength and hardness comparison of experimental findings with Junjin [1].

SEM of Alloy

The SEM sample structure of variable percentages indicated in Table **2** as alloys 1 and 2 are presented in Fig. (**6a** and **6b**), respectively. It is observed that the grain structures of as-cast alloy 2 contain equiaxed coarse grains. The twin's deformation is visible due to low dislocation density inside the grains, leading to larger grain size and reduced hardness. For magnesium alloy 1, the deformation twins start to reduce as the grains start to elongate. It is also observed that high angle grain boundaries start to originate. This indicates a greater hardness and tensile strength in Mg alloy composition 2 compared to 1.

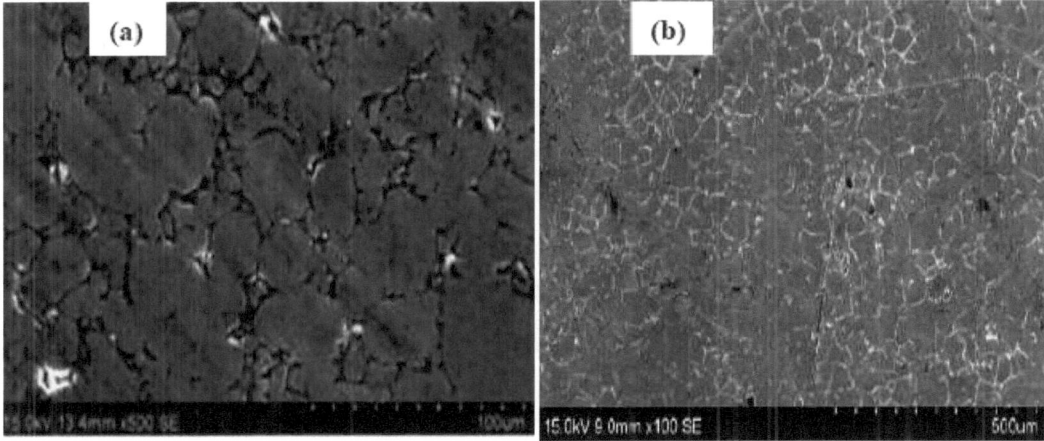

Fig. (6). SEM of **(a)** alloy **(b)** alloy 2.

CONCLUSIONS

The effects of Zn and Ca extracted on the mechanical properties of Al alloy were studied. The addition of the Zn element into aluminum alloys improved both the tensile strength and the corrosion resistance successfully; likewise, calcium improved the mechanical properties and corrosion resistance of the aluminum alloys. When 0.23% of organically extracted Zn and 0.23% of organically extracted calcium were added to the alloy, the ultimate tensile strength was 208.3 MPa. Comparing this to the reference experiment, which was 205 MPa. The following were observed in experimental results:

1. The addition of organically extracted Zn had a greater effect on refining grain, which could improve the mechanical properties of the as-cast alloys.
2. The presence of organically extracted Zn and Ca improved the ductility property of the alloy.

RECOMMENDATION

There is a need for further investigation of more cost effective, readily available, and organic sources of zinc and calcium that can be used for the reinforcement of magnesium alloy in future studies. This will enable the development of a cost effective Mg alloy.

CONSENT FOR PUBLICATION

Not Applicable.

CONFLICT OF INTEREST

The authors declare no conflict of interest, financial or otherwise.

ACKNOWLEDGEMENTS

Declared none.

REFERENCES

[1]   F. Junjin, "Mechanical properties of magnesium-alluminum alloy with the addition of zinc and calcium", *Mater. Res. Innov.,* vol. S4, pp. 228-234, 2017.

[2]   J. Chen, L. Tan, and K. Yang, "Recent advances on the development of biodegradable magnesium alloys: a review", *Mater. Technol.,* vol. 31, no. 12, pp. 681-688, 2016.
[http://dx.doi.org/10.1080/10667857.2016.1212587]

[3]   U. Riaz, L. Rakesh, I. Shabib, and W. Haider, "Effect of dissolution of magnesium alloy AZ31 on the rheological properties of Phosphate Buffer Saline", *J. Mech. Behav. Biomed. Mater.,* vol. 85, pp. 201-208, 2018.
[http://dx.doi.org/10.1016/j.jmbbm.2018.06.002] [PMID: 29908488]

[4]   J. Walker, S. Shadanbaz, T.B.F. Woodfield, M.P. Staiger, and G.J. Dias, "Magnesium biomaterials for orthopedic application: a review from a biological perspective", *J. Biomed. Mater. Res. B Appl. Biomater.,* vol. 102, no. 6, pp. 1316-1331, 2014.
[http://dx.doi.org/10.1002/jbm.b.33113] [PMID: 24458999]

[5]   L. Pompa, Z.U. Rahman, E. Munoz, and W. Haider, "Surface characterization and cytotoxicity response of biodegradable magnesium alloys", *Mater. Sci. Eng. C,* vol. 49, pp. 761-768, 2015.
[http://dx.doi.org/10.1016/j.msec.2015.01.017] [PMID: 25687006]

[6]   N.T. Kirkland, "Magnesium biomaterials: Past present and future", *Corros. Eng. Sci. Technol.,* vol. 47, pp. 322-328, 2012.
[http://dx.doi.org/10.1179/1743278212Y.0000000034]

[7]   D. Gastaldi, V. Sassi, L. Petrini, M. Vedani, S. Trasatti, and F. Migliavacca, "Continuum damage model for bioresorbable magnesium alloy devices - Application to coronary stents", *J. Mech. Behav. Biomed. Mater.,* vol. 4, no. 3, pp. 352-365, 2011.
[http://dx.doi.org/10.1016/j.jmbbm.2010.11.003] [PMID: 21316623]

[8]   P. Barlis, J. Tanigawa, and C. Di Mario, "Coronary bioabsorbable magnesium stent: 15-month intravascular ultrasound and optical coherence tomography findings", *Eur. Heart J.,* vol. 28, no. 19, pp. 2319-2330, 2007.
[http://dx.doi.org/10.1093/eurheartj/ehm119] [PMID: 17485484]

[9]   C. Riccardo, and V. Maurizio, "Metal Matrix Composites Reinforced by Nano-Particles", *Metals (Basel),* vol. 2, no. 3, pp. 65-83, 2014.

[10]   B. Zberg, P.J. Uggowitzer, and J.F. Löffler, "MgZnCa glasses without clinically observable hydrogen evolution for biodegradable implants", *Nat. Mater.,* vol. 8, no. 11, pp. 887-891, 2009.
[http://dx.doi.org/10.1038/nmat2542] [PMID: 19783982]

[11]   Z. Erlin, and Y. Lei, "A Microstructure, mechanical properties and bio-corrosion properties of Mg-Z--Mn-Ca alloy for biomedical application", *Mater. Sci. Eng.,* vol. 5, no. 7, pp. 212-218, 2018.

CHAPTER 5

Optimisation of Surface Quality, Process Conditions, and Characterisation of Additive Manufactured Components

Moses Oyesola[1,*], Ilesanmi Daniyan[1], Khumbulani Mpofu[1], Ntombi Mathe[2] and Lerato Tshabalala[2]

[1] *Department of Industrial Engineering, Tshwane University of Technology, Pretoria 0001, South Africa*

[2] *Laser Enabled Manufacturing, National Laser Centre, Council for Scientific and Industrial Research, Pretoria, South Africa*

Abstract: Additive manufacturing (AM) is widely known as a method of manufacturing components and parts from powder or wire elements emanating from layer by layer processing. Surface quality is a critical characteristic of any product manufactured additively. Hence, this study presents a process of selective laser melting used to manufacture 10 mm thin-walled metal tubes in cubes from Ti6Al4V powders. The laser used was an IPG YLS 500 Ytterbium fibre laser operating at 1076 nm wavelength with a 50 µm fibre delivery system using different laser contour scanning parameters. The scanner used was an Intelliweld 30 FC V system. A custom-built selective laser-melting platform enclosed within an inert glovebox enclosure was used for the part building. A complementary surface engineering strategy was employed using the statistical model approach of response surface methodology (RSM) to analyse the surface finish quality of SLM fabricated Ti6Al4V alloy. The selected variables optimised were the power density and consolidation rates, with their interactive effect on the experimental responses (surface roughness and top edge quality). The results obtained indicated that higher consolidation rates and mid-range power densities had better surface finishes due to a more stable melt pool. The findings of this work will add to the understanding of the process design and optimisation of components manufactured additively in order to promote the integrity of the developed product.

Keywords: Additive manufacturing, RSM, Selective laser melting, Surface quality.

* **Corresponding Author Moses Oyesola:** Department of Industrial Engineering, Tshwane University of Technology, Pretoria 0001, South Africa;Tel: +27840276799; E-mail: OyesolaMO@tut.ac.za

Ilesanmi Afolabi Daniyan (Ed.)

INTRODUCTION

Additive Manufacturing (AM) deals with the creation of objects with precise geometric shapes *via* Computer Aided Design (CAD), creation of 3D objects, and subsequent deposition layer by layer to make the final product [1, 2]. It is a manufacturing process whereby the final product can be produced directly from 3D models [3, 4]. This is in contrast to the traditional manufacturing process that is subtractive in order to machine a material to the required finish. The common AM techniques include Selective Laser Sintering (SLS), Fused Deposition Modelling (FDM), Fused Filament Fabrication (FFF), stereolithography, *etc.* [5 - 7]. AM boasts of merits such as manufacturing of products with complex or intricate shapes, reduction of inventory cost, elimination of complex tooling, manufacturing time, and cost effectiveness [8 - 10]. Previous studies have established that the AM offers a realistic possibility of developing products in a sustainable process in terms of material conservation, energy utilization, cost and environmental friendliness [11, 12]. The benefit of design freedom in AM often makes possible product's customisation as well as the development of product's variety simple to complex products [13, 14]. However, the demerits of the process include part selection, its non-suitability for high volume production, as well as the need for post-processing [15, 16]. It is noteworthy that the AM process is still emerging and that not all the components or parts of an assembly or system can be manufactured additively. This is coupled with the fact that the possibility of product development through AM for some emerging materials poses some manufacturing constraints. There is a need for the identification and selection of parts that can be developed *via* the process of additive manufacturing. This will ensure that the product meets the required service and functional requirements. The part to be developed *via* the AM process and the nature of the material to be employed largely determine the AM technique to be used. The AM is suitable for low volume production, which may affect the economies of scale. Furthermore, the finish of products manufactured through the AM process often requires post processing with time and cost implications. However, many efforts have been made in the area of process design and optimization in order to ensure the surface integrity of products made *via* the AM process [17 - 19]. This is aimed at eliminating the need for post-processing and ensuring the quality of the final product. However, due to the emerging materials and the dynamics of the various AM processes, there is still a dearth of information regarding the process optimisation of components manufactured additively. The aim of this study is to investigate the top edge quality of products manufactured using titanium (Ti6Al4V) powders and minimise the surface roughness. The two process parameters considered were the power density and consolidation rates. This is due to the fact that many works have been reported on the effect of other process parameters such as laser power, scan speed, and spot size on the surface finish of components manufactured *via* the AM process. The findings of this work will add to the understanding of the process design and optimisation of components manufactured additively in order to promote the integrity of the developed product. The understanding of process design issues of the AM processes is necessary in order to improve the capabilities of the materials used as well as hardware and products manufactured. The succeeding sections present the methodology employed, the results and discussion of the findings, as well as the conclusion and recommendations.

MATERIALS AND METHOD

Physical Experimentation

The experiments were set up on the custom-built selective laser-melting platform within an inert glovebox enclosure. The laser used was a 5 kW IPG YLS 5000 ytterbium fibre laser with a wavelength of 1076 nm and a delivery fibre core diameter of 50 μm. The scanner used was an Intelliweld 30 FC V system. Materials used were Ti6Al4V, gas atomized with the particle size distribution of 20 - 60 μm supplied by TLS Technik GmbH & Co. (Germany). The standard Aeroswift operating parameters for spot size and layer thickness were unchanged throughout the experiments. The samples built were 10 mm x 10 mm x 15 mm hollow square thin-walled metal in cubes to investigate the effect of power density and consolidation rate on surface finish. The tests were repeated three times. The chemical analysis provided by the powder is presented in Table **1**, while Fig. (**1**) shows the image of the custom designed and built glovebox.

Table 1. Material morphology of the powders used for the experiments.

Element	wt. % (max.)
Al	6.5
V	4.5
Fe	0.25
Si	NA
O	0.13
C	0.08
N	0.05
H	0.012
Ti	Remaining

Fig. (1). Image of the custom designed and built glovebox.

Numerical Experimentation

The Design Expert software was employed for modelling and further analysis. The aim is to obtain the optimal response of the inputs to the output through a quadratic model. The choice of RSM stems from the fact that the technique is suitable for the optimisation of the process parameter [20, 21]. This design consists of a complete 2k factorial design, where k is the number of variables whose factors levels are coded as -1 and 1. The factors and levels used in the factorial design are given in Table **2**. The range of the two process parameters were: power density (3 - 6.5 W/m^3) and consolidation rates (50 - 175 mm/s). The interaction time ranges between 50 - 175 sec although, this was not included in the design matrix as a variable.

Table 2. Factors and levels used in a factorial design.

Parameters	Units	Min	Max	Coded Low	Coded High	Mean	Std. Dev.
Power density (A)	W/m^3	3	6.5	-1 â†" 3.00	+1 â†" 6.50	4.75	1.212
Consolidation rates (B)	mm/s	50	175	-1 â†" 50.0	+1 â†" 175.0	112.5	43.30

Design Expert provides prediction equations in terms of actual units and coded units. The coded equations are determined first, and the actual equations are derived from the code. To get the actual equation, each term in the coded equation is replaced with its coding formula according to Equation (1).

$$X_{coded} = \frac{X_{actual} - \bar{X}}{(X_{Hi} - X_{Low})/2}$$

(1)

where X_{coded} is the coded score for X_{actual} is the actual scores, and X bar is the assumed mean, while X_{Hi} and X_{Low} is the deviation of the values from the assumed mean.

The combination of the range of the process parameters, namely power density and interaction time, using the RSM, produced 36 experimental trials whose response (surface roughness) was determined *via* the physical experimentations. The spot size was kept constant throughout the experiments. The samples built were 10 x 10 x 15 (H) mm hollow square thin-walled cubes to investigate the effect of the laser parameters, power density, and consolidation rate on surface finish. The tests were repeated three times.

Power density (p_d) is given by Equation (2) while Equation (3) expresses the consolidation rate (C_r).

$$\rho_d = \frac{L_p-}{A} \tag{2}$$

where: L_p is the laser power (kW) and A is the spot area (mm^2)

$$C_r = S_z \times S_s \times t \tag{3}$$

where: S_z is the scan speed $(mm/s^2)S_s$ is the spot size (mm), and t is the thickness of layers.

Modelling of Surface Roughness

Controlling the surface roughness of SLM fabricated samples is of utmost importance, and numerous measures on how to solve the challenge of surface roughness have been proposed [22 - 24]. In line with the previous observations in the field of AM using SLM method, it is desirable to minimise the number of facets and reduce computational demands. Process parameters, the laser scan path, and the powder morphology introduce manufacturing error leading to roughness. Hence, roughness refers to a geometric deviation from a reference datum as denoted in Equation (4).

$$R_a = \frac{1}{n}\sum_{i=1}^{n}|y_j| \tag{4}$$

Where y_j is the vertical distance from the mean line to the jth data point and i is the facet of interest. To achieve minimal surface roughness outcomes by accommodating other observations such as surface inclination and aligning fabricated components patterns according to spatial rotations, an average roughness is specified in Equation (5).

$$R_{a\_ave} = \frac{\sum_{i=1}^{n} R_{a,i} \, \phi \, A}{\sum_{i=1}^{n} A} \tag{5}$$

Average roughness, R_{a_ave} refers to the area-weighted average roughness of each facet within the component geometry. The area, A, is assessed by the Cartesian coordinates of the associated vertices (Equation 6) or by Heron's rule with reference to the facet side lengths (a, b, and c) (Equation 7).

$$A = \frac{1}{2} | x_1 y_2 - x_1 y_3 + x_2 y_3 - x_2 y_1 + x_3 y_1 - x_3 y_2 | \tag{6}$$

$$\sqrt{A} = \sqrt{1/16[(a+b+c)(-a+b+c)(a-b+c)(a+b-c)]} \tag{7}$$

Weighted roughness of relevant surfaces, $R_{a\_w\_r}$, applies a roughness weighting $\lambda(\emptyset)$ to the measured surface roughness of each facet to promote the desired surface roughness distribution in the finished component (Equation 8).

$$R_{a\_w\_r} = \frac{\sum_{i=1}^{n} r_i \lambda(\phi) R_a(\phi) A_i}{\sum_{i=1}^{n} A_i r_i} \tag{8}$$

where R_{a_r}, facets are defined as either relevant ($r_i = 1$) or irrelevant ($r_i = 0$) to the associated surface roughness to avoid high roughness values for relevant surfaces.

The surface roughness of the samples was measured using a Mahrsurf PS1 profilometer. An average peak profile was calculated (Ra). Ra is defined as the arithmetic average of the roughness profile. An average reading of 3 measurements was taken to ensure the quality of the final product due to emerging materials. All Powder Bed Fusion (PBF), specifically Selective Laser Melting (SLM) systems operate by repeated melting of sequentially added layers; delamination often occurs when the interlayer bonding strength is exceeded by thermally induced residual stresses or incomplete melting, grain boundary precipitate formation, or other mechanisms. In delamination occurrence, studies have identified the importance of applied energy input and scan strategy; melt balls result in fabrication failure due to interference with the rake and non-uniform layer formation. Thus, delamination can be caused by either incomplete melting of powder feedstock or insufficient remelting of underlying layers to achieve high inter-layer strength. The factors affecting performance measures are not limited to interlayer bonding, surface finish, and tensile strength, and analysis of these factors is not limited to environmental control, powder raking or density, and control over the thermal history of a component. Hence, data collection and analysis is a necessary aspect of the fabrication process due to the possibility of correlating observational information with performance characteristics to make a balanced justification.

RESULTS AND DISCUSSION

Provided Table **2** presents the factors and levels used in the factorial design while Table **3** presents the feasible combination of the process parameters, namely the power density and consolidation rates as well as quality of the top edge of the product samples. The surface roughness obtained for samples 5, 6, 9, 15, 19, 21, 25, 26, and 30 were also reported. Sample 25 produced the least surface roughness (12.64 μm) compared to other samples, and the combination of the process parameters which produced this minimal surface roughness was: power density (5.8 kW/mm^2) and consolidation rate (20 mm/s). Delamination is a condition in which there is the separation of the layers of the samples developed *via* the AM process [25]. This phenomenon was observed in samples 28, 29, 30, 34, 35, and 35, while other samples did not show any sign of delamination. Delamination can affect the product functionality and promote the failure of such components under the required service conditions. Delamination implies the non-adherence of the layers due to incomplete melting due to insufficient melting temperature. On the other hand, when the melting temperature exceeds the optimum, a combination of melt pool size and surface tension may result in swelling or melt balling. Hence, the need to tightly control the range and variations of the process parameters in order to avoid these defects. In addition, the top edge quality of samples 4, 12, 14, 16, 17, 18, 22, 23, 24, 27, 32 and 33 were poor. The poor quality of the top edge of the samples can be traced to the combination of the process parameters, namely power density and consolidation rates, which were not found within the optimal range. Hence, the need for optimisation to keep the process parameters within the optimum range.

Table 3. Laser bulk parameters used for the experiments.

Sample #	Power density (kW/mm^3)	Consolidation rate (mm/s)	Surface roughness Ra (μm)	Delamination	Top edge quality
1	3	20.00	-	No	Ok
2	3	13.33	-	No	Ok
3	3	10.00	-	No	Ok
4	3	8.00	-	No	Poor

(Table 3) cont.....

Sample #	Power density (kW/mm³)	Consolidation rate (mm/s)	Surface roughness Ra (μm)	Delamination	Top edge quality
5	3	6.67	14.77	No	Ok
6	3	5.71	13.62	No	Ok
7	3.7	20.00	-	No	Ok
8	3.7	13.33	-	No	Ok
9	3.7	10.00	13.52	No	Ok
10	3.7	8.00	-	No	Ok
11	3.7	6.67	-	No	Ok
12	3.7	5.71	-	No	Poor
13	4.4	20.00	-	No	Ok
14	4.4	13.33	-	No	Poor
15	4.4	10.00	13.61	No	Ok
16	4.4	8.00	-	No	Poor
17	4.4	6.67	-	No	Poor
18	4.4	5.71	-	No	Poor
19	5.1	20.00	13.53	No	Ok
20	5.1	13.33	-	No	Ok
21	5.1	10.00	17.75	No	Ok
22	5.1	8.00	-	No	Poor
23	5.1	6.67	-	No	Poor
24	5.1	5.71	-	No	Poor
25	5.8	20.00	12.64	No	Ok
26	5.8	13.33	14.45	No	Ok
27	5.8	10.00	-	No	Poor
28	5.8	8.00	-	Yes	N/A
29	5.8	6.67	-	Yes	N/A
30	5.8	5.71	14.07	Yes	N/A
31	6.5	20.00	-	No	Ok
32	6.5	13.33	-	No	Poor
33	6.5	10.00	-	No	Poor
34	6.5	8.00	-	Yes	N/A
35	6.5	6.67	-	Yes	N/A
36	6.5	5.71	-	Yes	N/A

Effect of Laser Bulk Parameters on Surface Roughness

Fig. (**2a** and **2b**) show the sample tubes as built. The samples 28, 29, 30 and 34, 35, 36 were of poor quality and delaminated before the build was complete (Fig. **2**). These samples typically had lower consolidation rates, interaction time, and higher power density. This can indicate an unstable melt-pool and a large degree of spatter due to the higher powers and lower speeds; the higher the interaction time, the higher the rate of consolidation and *vice versa*. The consolidation rate is a function of the scan speed, spot size, and thickness of layers. Under a high rate of consolidation, it implies that there is sufficient time for the powders to melt together, thereby resulting in the complete adherence of the layers. In the performance evaluation for this investigation, the first step involved the visual inspection of the thin-walled cubes to identify delamination or related defects. The subsequent step was the investigation of the top edge quality of the product samples, as presented in Fig. (**2**). While Fig. (**3**) shows an example of a poor edge quality (depicted by arrows). It can be seen that sample 22 in Fig. (**3**) has a jagged and rough looking top edge, whereas sample 25 has a smooth looking top edge.

Fig. (2). As built titanium cube (**a**) experimental samples and (**b**) the top edge quality.

Fig. (3). As built samples showing the top edge quality.

Lower overlaps or increased hatch contour spacings produced better surface finish; however, the porosity just below the surface increases with the low overlap. The implementation of the contour scans significantly improved surface finish regardless of the degree of overlap. Density comparisons between micro-CT and Archimedes show that due to surface effects on Archimedes and specific regions selected on the micro-CT system, micro-CT shows higher density results than the Archimedes results. Porosity on the Archimedes technique is lower due to surface effects. Roughness results are higher on the micro-CT method due to its resolution limits and detection of deep open surface porosity defects that cannot be picked up by the conventional mechanical tactile probes.

RESULTS FROM THE NUMERICAL EXPERIMENTATION

The interaction response surface plots are the graphical representation useful to understand interaction properties between the input and output parameters. The ultimate aim of the plot is to predict the optimum values of the variables such the responsesare maximized or minimised. From the analysis of the interaction graphs, the major parameter that influences wall thickness is the powder density. Fig. (**4** and **5**) present the effect of the interaction time and power density on the wall thickness of the product samples. The higher the interaction time, the higher the rate of consolidation and the lower the chances for delamination and *vice versa*. The results obtained indicated that initially, the power density was insufficient, therefore resulting in delamination in some of the samples. An increase in the power density within the optimum range promotes a good surface finish. An increase in the magnitude of the powder density beyond this optimum range may result in an increase in the size of the melt pool and surface tension, which may result in swelling or melt balling with attendant poor surface finish.

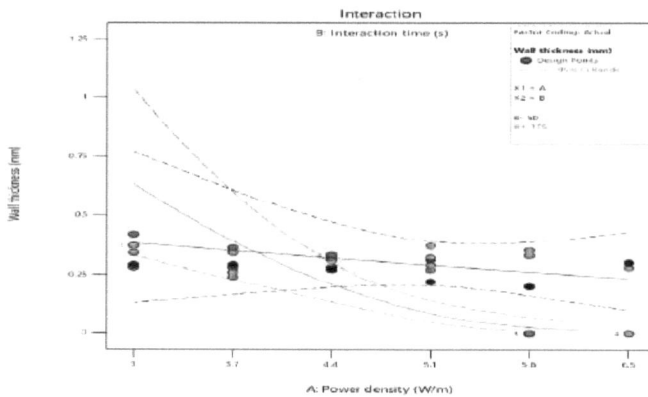

Fig. (4). Response surface plot for wall thickness.

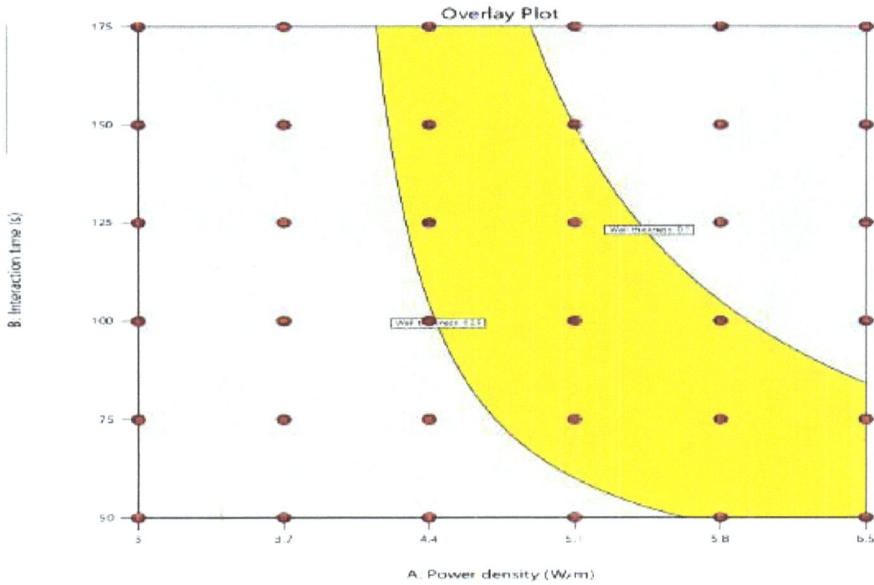

Fig. (5). Response surface contour plot for wall thickness.

These samples were also subject to roughness measurements performed on the Mahrsurf PS1 profilometer. The results indicate that the surface finish improves with an increase in the consolidation rates up to 20 mm/s and power density up to 5.8 kw/m^3. There are many theories on the effects of the process parameters on surface finish. According to Calignano *et al.* [26], a low scan speed could increase the volume of liquid produced within the melt pool. This tends to widen the same melt pool provoking a larger thermal difference across it and consequently a greater variation in surface tensions and finish. An attempt to reduce these changes may cause the melt pool to break off into smaller entities, well known as "balling," which solidify at the edge of the melt pool, thus, increasing the surface roughness. On the other hand, higher speeds may also lead to smaller and narrower melt pools, thus giving favourable surface finishes. In addition, larger melt pools can drag semi-molten particles from adjacent areas into the melt pool and become embedded in the edge of the track surface, thereby contributing to an increase in roughness. The trend generally shows that samples with higher consolidation rates and mid-range power densities have better surface finishes due to a more stable melt pool. From these test samples, 25 parameter set was selected for the subsequent testing. Fig. (6) shows the surface profile for sample 25, which has the least surface roughness.

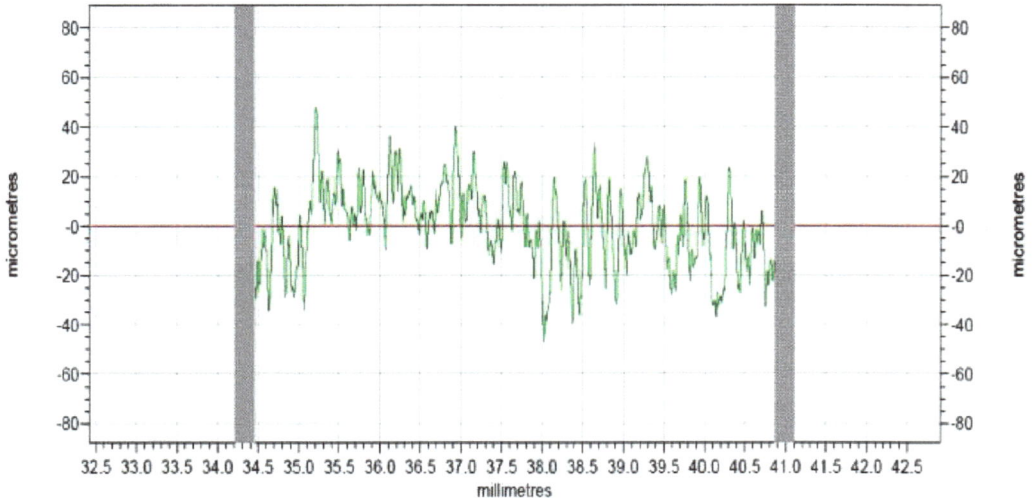

Fig. (6). Roughness profile of sample 25.

CONCLUSION

A contour scanning strategy varying laser power density and laser consolidation rates was utilised as a method to improve the surface finish of as-built AM samples. The first step of the experiments was to test a matrix of power densities and consolidation rates using contour scans only (hollow tubes). These parameter sets were inspected visually for delamination and top edge surface quality. The selected samples were then eliminated based on a light test for porosity indication. Based on this elimination criteria, the selected samples were scanned for surface roughness *via* a Mahrsurf PS1 mechanical roughness probe for Ra measurements. The trend generally shows that samples with higher consolidation rates and mid-range power densities have better surface finishes due to a more stable melt pool. The best parameter was taken from the first study to investigate the effect of hatch contour overlap on the surface finish with the use of micro-CT and mechanical profilometry. 10 mm cubic samples were built with a standard hatch parameter and contour parameter from sample 25. Lower overlaps or increased hatch contour spacings produced better surface finishes; however, the porosity just below the surface increases with the low overlap. The implementation of the contour scans significantly improved surface finish regardless of the degree of overlap. Density comparisons between micro-CT and Archimedes show that due to surface effects on Archimedes and specific regions selected on the micro-CT system, micro-CT shows higher density results than the Archimedes results. Porosity on the Archimedes technique is lower due to surface effects. Roughness results are higher on the micro-CT method due to its resolution limits and detection of deep open surface porosity defects that cannot be picked up by the conventional mechanical tactile probes. Future works can consider the comparative analysis of the surface roughness and top edge quality of the same material under the same conditions for the additive and subtractive manufacturing processes.

CONSENT FOR PUBLICATION

Not applicable.

CONFLICT OF INTEREST

The authors declare no conflict of interest, financial or otherwise.

ACKNOWLEDGEMENTS

Declared none.

REFERENCES

[1]   B.C. Gross, J.L. Erkal, S.Y. Lockwood, C. Chen, and D.M. Spence, "Evaluation of 3D printing and its potential impact on biotechnology and the chemical sciences", *Anal. Chem.*, vol. 86, no. 7, pp. 3240-3253, 2014.
[http://dx.doi.org/10.1021/ac403397r] [PMID: 24432804]

[2]   F. Regina, F. Lavecchia, and L.M. Galantucci, "Preliminary study for a full colour low cost open source 3D printer, based on the combination of fused deposition modelling (FDM) or fused filament fabrication (FFF) and inkjet printing", *Int. J. Interact. Des. Manuf.*, vol. 12, pp. 979-993, 2018.
[http://dx.doi.org/10.1007/s12008-017-0432-x]

[3]   R. Huang, M. Riddle, D. Graziano, J. Warren, S. Das, S. Nimbalkar, and E. Masanet, "Energy and emissions saving potential of additive manufacturing: the case of lightweight aircraft components", *J. Clean. Prod.*, vol. 135, pp. 1559-1570, 2016.
[http://dx.doi.org/10.1016/j.jclepro.2015.04.109]

[4]   W. Gao, Y. Zhang, D. Ramanujan, K. Ramani, Y. Chen, C.B. Williams, and P.D. Zavattieri, "The status, challenges, and future of additive manufacturing in engineering", *Comput. Aided Des.*, vol. 69, pp. 65-89, 2015.
[http://dx.doi.org/10.1016/j.cad.2015.04.001]

[5]   I.A. Daniyan, V. Balogun, K. Mpofu, and F.T. Omigbodun, "An interactive approach towards the development of an additive manufacturing technology for railcar manufacturing", *International Journal on Interactive Design and Manufacturing*, vol. 14, pp. 651-666, 2020.
[http://dx.doi.org/10.1007/s12008-020-00659-8]

[6]   M.S. Hossain, J.A. Gonzalez, R.M. Hernandez, M.A.I. Shuvo, J. Mireles, A. Choudhuri, and R.B. Wicker, "Fabrication of smart parts using powder bed fusion additive manufacturing technology", *Addit. Manuf.*, vol. 10, pp. 58-66, 2016.
[http://dx.doi.org/10.1016/j.addma.2016.01.001]

[7]   M. Brandt, "S.J., Sun, M. Leary, S. Feih, J. Elambasseril and Q.C. Liu. "High-value SLM aerospace components: from design to manufacture", *Adv. Mat. Res.*, vol. 633, pp. 135-147, 2013.

[8]   A. Salea, R. Prathumwan, and J. Junpha, "J. and K. Subannajui. "Metal oxide semiconductor 3D printing: preparation of copper (II) oxide by fused deposition modelling for multi-functional semiconducting applications", *J. Mater. Chem. C Mater. Opt. Electron. Devices*, vol. 5, no. 19, pp. 4614-4620, 2017.

[http://dx.doi.org/10.1039/C7TC00990A]

[9]　C. Zhou, Y. Chen, and Z. Yang, "Digital material fabrication using mask-image-projection-based stereolithography", *Rapid Prototyp,* vol. 19, no. 3, pp. 153-165, 2013.

[10]　I.A. Daniyan, K. Mpofu, O.L. Daniyan, F. Fameso, and M.O. Oyesola, "Computer aided simulation and performance evaluation of additive manufacturing technology for component parts manufacturing", *Int. J. Adv. Manuf. Technol.,* vol. 107, pp. 4517-4530, 2020.
[http://dx.doi.org/10.1007/s00170-020-05340-8]

[11]　D. Wang, W. Dou, and Y. Yang, "Research on selective laser melting of Ti6Al4V: Surface morphologies, optimized processing zone, and ductility improvement mechanism", *Metals (Basel),* vol. 8, no. 7, p. 471, 2018.
[http://dx.doi.org/10.3390/met8070471]

[12]　M.O. Oyesola, K. Mpofu, N.R. Mathe, and I.A. Daniyan, "Hybrid-additive manufacturing cost model: a sustainable through-life engineering support model for maintenance repair overhaul in the aerospace", *Procedia Manuf.,* vol. 49, pp. 199-205, 2020.
[http://dx.doi.org/10.1016/j.promfg.2020.07.019]

[13]　N. Knofius, M.C. van der Heijden, and W.H. Zijm, "Consolidating spare parts for asset maintenance with additive manufacturing", *Int. J. Prod. Econ.,* vol. 208, pp. 269-280, 2019.
[http://dx.doi.org/10.1016/j.ijpe.2018.11.007]

[14]　L.N. Carter, and K. Essa, "K. and M.M. Attallah. "Optimisation of selective laser melting for a high temperature Ni-superalloy", *Rapid Prototyping J.,* vol. 21, no. 4, pp. 423-432, 2015.
[http://dx.doi.org/10.1108/RPJ-06-2013-0063]

[15]　J. Wang, H. Xie, Z. Weng, T. Senthil, and L. Wu, "L. "A novel approach to improve mechanical properties of parts fabricated by fused deposition modeling", *Mater. Des.,* vol. 105, pp. 152-159, 2016.
[http://dx.doi.org/10.1016/j.matdes.2016.05.078]

[16]　D. Chaidas, K. Kitsakis, J. Kechagias, and S. Maropoulos, "The impact of temperature changing on surface roughness of FFF process", *IOP Conf Ser Mater Sci Eng,* vol. vol. 161, 2016pp. 1-9
[http://dx.doi.org/10.1088/1757-899X/161/1/012033]

[17]　I.A. Daniyan, K. Mpofu, M. O. Oyesola, and L. Daniyan, "Process optimization of additive manufacturing technology: a case evaluation of a manufactured railcar accessory", *Procedia CIRP,* vol. 95, pp. 89-96, 2020.
[http://dx.doi.org/10.1016/j.procir.2020.01.143]

[18]　A.I. Khuri, and A.J. Cornell, "Response surfaces, designs and analyses, revised and expanded, chapter 2, matrix algebra, least squares, the analysis of variance, and principles of experimental design", In: *Marcel Dekker, Inc,* 1996.

[19]　L.-I Tong, C.-H Wang, and H.-C Chen, "Optimization of multiple responses using principal component analysis and technique for order preference by similarity to ideal solution", *International Journal of Advanced Manufacturing Technology,* vol. 27, no. 3-4, pp. 407-414, 2005.

[20]　I.A. Daniyan, I. Tlhabadira, S.N. Phokobye, M. Siviwe, and K. Mpofu, "Modelling and optimization of the cutting forces during Ti6Al4V milling process using the Response Surface Methodology and dynamometer", *MM Science Journal,* vol. 128, pp. 3353-3363, 2019.
[http://dx.doi.org/10.17973/MMSJ.2019_11_2019093]

[21]　"Daniyan, I. Tlhabadira, K. Mpofu and A.O. Adeodu. "Development of numerical models for the prediction of temperature and surface roughness during the machining operation of titanium alloy (Ti6Al14V)", *Acta Polytechnica Journal,* vol. 60, no. 5, pp. 369-390, 2020.
[http://dx.doi.org/10.14311/AP.2020.60.0369]

[22]　Min-Ho Hong, Bong Ki Min, and Tae-Yub Kwon, "The influence of process parameters on the surface roughness of a 3D-Printed Co–Cr dental alloy produced via selective laser melting", *Appl. Sci,* vol. 6, no. 401, pp. 1-10, 2016.

[http://dx.doi.org/10.3390/app6120401]

[23]   Y. Pupo, K.P. Monroy, and J. Ciurana, "Influence of process parameters on surface quality of CoCrMo produced by selective laser melting", *Int. J. Adv. Manuf. Technol.,* vol. 80, pp. 985-995, 2015. [http://dx.doi.org/10.1007/s00170-015-7040-3]

[24]   C. Lei, J. Cui, Z. Xing, and H. Fu, *Phys. Procedia,* vol. 25, pp. 118-124, 2012. [http://dx.doi.org/10.1016/j.phpro.2012.03.059]

[25]   W.J.   Sames,   F.A.   List,   S.   Pannala,   and   R.R.   Dehoff,   "[Online}   Available   at https://www.osti.gov/pages   /servlets/purl   /1267051", https://www.osti.gov/pages/servlets/purl/1267051

[26]   F. Calignano, D. Manfredi, E.P Ambrosio, L. Iuliano, and P. Fino, "Influence of process pa-rameters on surface roughness of aluminium parts produced by DMLS", *Int. J. Adv. Manuf. Technol,* vol. 67, no. 9-12, pp. 2743-2751, 2013.

CHAPTER 6

Development and Optimisation of Additively Manufactured Radiometer Casing for Cosmic Particles Characterization

Lanre O. Daniyan[1,*], Matthew O. Afolabi[2], Ilesanmi A. Daniyan[2] and Felix Ale[3]

[1] *Department of Physics and Electronics, Adekunle Ajasin University, P. M. B. 0001, Akungba Akoko, Ondo State, Nigeria*

[2] *Department of Industrial Engineering, Tshwane University of Technology, Pretoria 0001, South Africa*

[3] *National Space Research and Development Agency (NASRDA), P.M.B. 437, Abuja, Nigeria*

Abstract: Additive manufacturing (AM) is a digital technology for producing components directly from a 3D model. This study develops the radiometer casing using the Fused Filament Fabrication (FFF) of the AM technology. The 3D model of the casing was modeled in a Rhinoceros environment while the implementation was carried out using the FFF. The optimisation of the process parameters was carried out using the Response Surface Methodology (RSM) and the Central Composite Design (CCD). The range of the process parameters were: extrusion temperature (230-250°C), extrusion velocity (50 mm/sec-250 mm/sec), filament orientation (0-90°) and layer thickness (0.10-0.50 mm). Taking the surface roughness as the response of the designed experiment, the statistical analysis of the results obtained from the numerical and physical experiments was used to obtain a predictive model for surface roughness. Furthermore, the combinations of the process parameters that produced the least surface roughness (2.05μm) were: extrusion temperature (240 °C), extrusion velocity (150 mm/sec), filament orientation (45°) as well as layer thickness (0.30 mm). This study provides an insight into the feasible range of process parameters that will enhance the surface finish of products developed using Polyethylene Terephthalate Glycol (PETG) filament.

Keywords: Additive Manufacturing, Digital Technology, Fused Filament Fabrication, Polyethylene Terephthalate Glycol (PETG) filament, Process Parameter, Surface Roughness.

* **Corresponding Author Lanre O. Daniyan:**Department of Physics and Electronics, Adekunle Ajasin University, P. M. B. 0001, Akungba Akoko, Ondo State, Nigeria; Tel: +2348058484254; E-mail: danomartins@hotmail.com

Ilesanmi Afolabi Daniyan (Ed.)

INTRODUCTION

Additive Manufacturing (AM) technology is an emerging digital manufacturing technology suitable for developing parts directly from a digital model [1]. AM offers a range of benefits, and as such, the technology finds application in the aerospace, biomedical, manufacturing, architecture fields, *etc.,* for components manufacturing [2 - 5]. Some of the benefits of the AM process include time and cost effectiveness during manufacturing, ease of customisation, elimination of costly tools and fixtures, freedom of design, suitability for rapid prototyping, reduction in manufacturing wastages, production of intricate geometries, and component manufacturing on a small scale [6 - 10]. The benefits of the AM technology are taken into account for the development of the radiometer casing in this study. However, some challenges of the AM process have been widely reported. These include surface roughness and the need for post processing, variation of process parameters, variation in the various stages of the AM processes, *etc.* [11 - 14]. Some of the approaches employed by several researchers to solve the identified challenges include process optimisation, modelling and simulation, process monitoring, and control techniques, amongst others [15 - 20]. Daniyan *et al.* [16] reported the process optimisation of additive manufacturing technology using a manufactured rail accessory as a case evaluation. The study provides a basis for preliminary process design as well as the determination of the feasible process parameters that can assist in meeting the finish requirements of a product. Natha *et al.* [11] carried out the optimisation of process parameters for fused filament fabrication process parameters in order to achieve dimensional accuracy of part geometry. The developed model demonstrated capacity for process design optimisation under uncertainty in minimising errors related to layer thickness and printing time. Weake *et al.* [12] used the Taguchi method for the optimisation of the process parameters of Fused Filament Fabrication in order to achieve optimum tensile strength for the component samples fabricated using Acrylonitrile Butadiene Styrene (ABS) material. The results obtained indicate that the variation of process parameters significantly influences the mechanical properties of the developed components. Hong *et al.* [21] and Pupo *et al.* [22] also reported that the variations in the magnitude of the process parameters could influence the surface finish of additively manufactured components. The aim of this study is to develop the radiometer casing using Polyethylene Terephthalate Glycol (PETG) filament and optimise the process parameters.

The novelty of this study lies in the fact that the process optimisation of process parameters for Polyethylene Terephthalate Glycol (PETG) filament has not been sufficiently highlighted by the existing literature. The findings of this study will assist manufacturers indetermining the optimum range of the process parameters when using PETG filament material for product development. The succeeding sections present the materials and method, results and discussion, as well as the conclusion and recommendations.

MATERIALS AND METHOD

The methodology basically comprises numerical and physical experimentation techniques. Before the development of the prototype, the combination of both the numerical and physical experimentations was used to obtain the feasible range of the process parameters.

Numerical Experimentation

The numerical technique involved the use of Response Surface Methodology (RSM) and the Central Composite Design (CCD). The RSM and CCD are viable techniques for determining the feasible combinations of process parameters and their cross effects on the response of the designed experiment [23 - 27]. Furthermore, the technique is suitable for process modelling and optimisation for obtaining improved product's quality as well as the cost and time effectiveness of the manufacturing process [28]. As shown in Table **1**, the Design of Experiment (DoE) consists of four factors A, B, C, and D (A: extrusion temperature, B: extrusion velocity, C: filament orientation, and D: Layer thickness). These are the dependent variables. The independent variable is the response of the designed experiment, which is the surface roughness. The DoE was implemented in the design expert (version 8 software) environment. The combination of the process parameters using RSM generated 13 experimental trials whose responses were determined through physical experimentations.

Table 1. Summary of the numerical experimentation.

Notation	Independent Variables	Levels		
		-1	0	1
A	Extrusion temperature (°C)	230	240	250
B	Extrusion velocity (mm/s)	50	150	250
C	Filament orientation (deg.)	0	45	90
D	Layer thickness (mm)	0.10	0.30	0.50

The summary of the DoE is presented in Table **1**.

The range of the process parameters namely: extrusion temperature (230-250°C), extrusion velocity (50 mm/sec-250 mm/sec), filament orientation (0-90°) and layer thickness (0.10-0.50 mm) were selected based on selected previous studies [11, 12, 16, 32].

The statistical analysis carried out involves the Analysis of Variance (ANOVA) for the validation of the developed model. The indicators of valid numerical experimentation include the "p-value Prob > F", which should be less than 0.050 to ensure that the model terms are statistically significant. Other indicators include the "Lack of Fit" which should be statistically insignificant relative to the pure error, as well as the correlation coefficients, namely the predicted R square, R squared, and the adjusted R squared, which are expected to be within the same range and close to 1 [29]. Also, the adequate precision, which measures the signal to noise ratio. A ratio greater than 4 is usually desirable for a significant model. A model with a small p-value which is less than 0.05 and a large F-value greater than unity, is usually considered significant [30].

Physical Experimentation

The 3D model of the casing for a radiometer was carried out using Rhinoceros, a CAD environment (Fig. **1**). Thereafter the slicing of the 3D model was done with the aid of the CURA slicer having a configuration of 0.1 mm layer height. The G-code was fed into Whanhao duplicator i3 plus for the Fused Filament Fabrication (FFF) of the casing. The Whanhao duplicator i3 plus is a single extruder system that has a build volume capacity of 200x200x180 mm [31]. The framework for the implementation of the FFF of the casing is presented in Fig. (**2**). A sky blue and white Polyethylene Terephthalate Glycol (PETG) filament was employed for the casing fabrication, as shown in Fig. (**3**). The choice of the PETG stems from the fact that the material is suitable for many industrial applications due to its excellent physical, thermal, electrical, and mechanical properties. Some of the properties of the PETG include high resistance to a temperature, which makes it suitable for high temperature applications, high strength, and excellent resistance to corrosion [32]. These properties are necessary for the service requirement of the final product (radiometer casing). Furthermore, the material is highly durable and boasts of good printability during the AM process. Good printability is linked to the quality of the final product, which is highly essential in order to ensure a good quality product that will meet the service requirements and also minimise reverse logistics [16]. Considering the choice of sustainable material in terms of sustainability, the PETG is also a good candidate. The material characteristically emits lesser fumes during the AM process, which contributes to a substantial reduction in environmental pollution [32]. In addition to this, the low bed temperature (about 60 °C - 80 °C) and printing temperature (about 230 °C-250 °C) [32] imply a significant reduction in energy consumption during the additive manufacturing process. The lower the energy consumed during the AM process, the more sustainable and environmentally friendly the process is and *vice versa* [33]. While the low temperature may reduce the rate of melting of the polymer, excessive temperature beyond the optimum value will increase the rate of energy consumption, thereby making it less sustainable. This could also cause the filament to burn out, thus resulting in dimensional variability of the final product [33]. The properties of the PETG employed are presented in Table **1**.

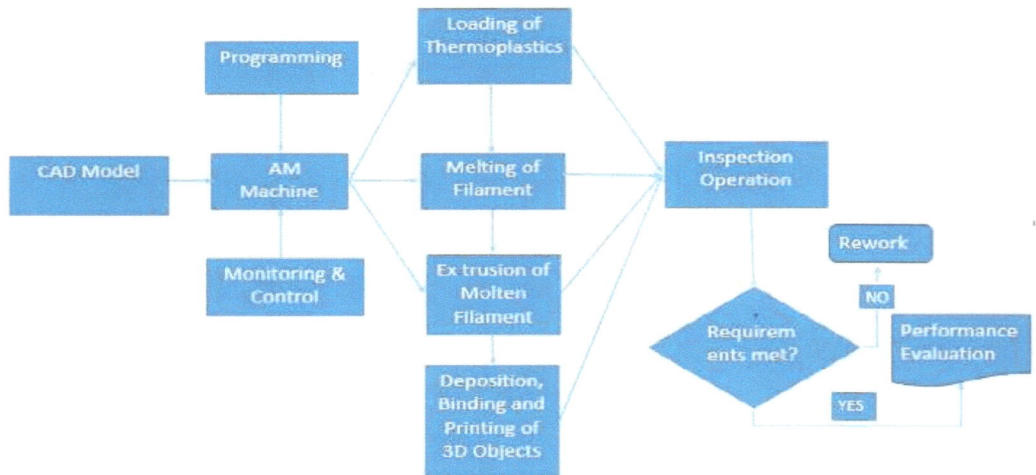

Fig. (1). Block diagram for the additive manufacturing process [23].

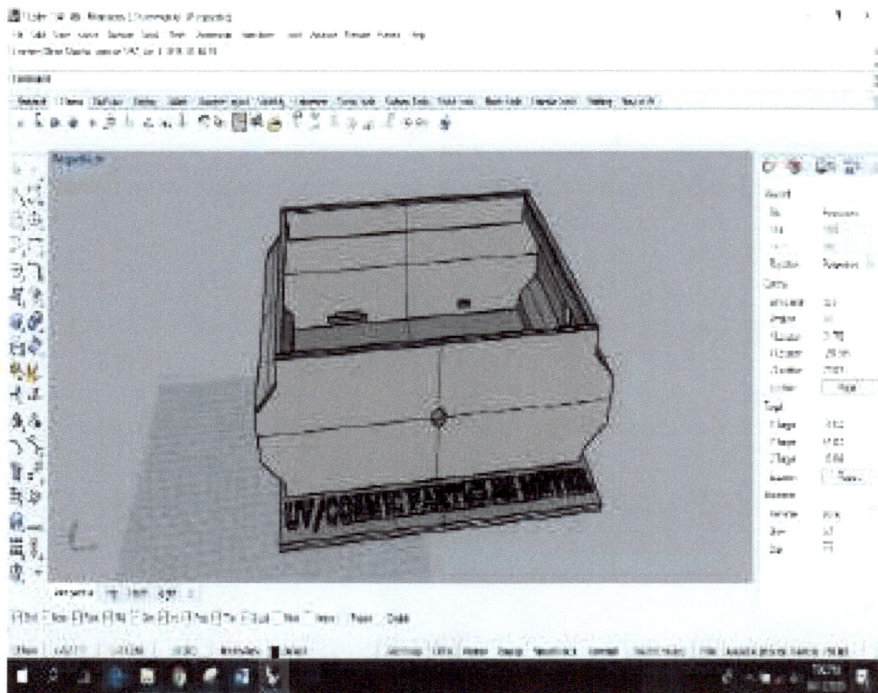

Fig. (2). 3D model of the casing Figure 3. The developed radiometer casing.

Table 2. PETG properties [32]

Properties	Value
Softening temperature (°C)	85
Heat deflection temperature (°C)	70
Impact strength (KJ/m^2)	11
Flexural modulus (MPa)	1880
Density (kg/m^3)	31270
Melt flow index (g/min)	11

The Fused Filament Fabrication is an extrusion based technique of additive manufacturing that involves a continuous extrusion and subsequent filament deposition from a heated nozzle [34]. Fig. (1) summarises the FFF process as follow; creation of a 3D CAD model, slicing of the 3D model into layers using the slicing software, feeding of thermoplastics and subsequent melting, extrusion of the molten plastics, deposition in thin layers, one on top of another and onto the print bed, binding and production of a 3D manufactured part [35]. FFF was considered for the development of the radiometer casing because the process boasts easy set-up, cost effectiveness, extensive freedom of design, fabrication of complex geometries, and suitability for polymer and plastic based materials [12].

Fig. (2) shows the 3D model of the casing in a Rhinoceros environment, while Fig. (3) shows the implementation using the FFF process of additive manufacturing.

The top and side surface roughness of the samples were measured using the Mitutoyo SJ – 201 surface roughness machine.

The surface and side roughness were measured according to EN ISO 4287 and EN ISO 16610-21 standards. The parameters considered during the roughness measurement are presented in Table 3 [36].

Table 3. Parameters considered for roughness measurement.

Parameters	Specification
Dimension of the sample (mm)	2.0×2.0×2.0
Maximum probe tip radius (μm)	2.00
Sampling length (mm)	0.08
Evaluation length (mm)	0.40
Stylus travel (mm)	0.48

(Table 3) cont.....

Parameters	Specification
Point pitch	Standardised in the measuring device
Cut-off wavelength	Standardised in the measuring device

RESULTS AND DISCUSSION

The results obtained from the numerical and physical experimentations, which show the feasible combination of process parameters as well as the resulting actual, and predicted values of the surface roughness, are presented in Table **4**. Table **5** presents the results obtained from the statistical analysis of the developed model.

Table 4. The process parameters and the resulting actual and predicted surface roughness.

Trials	Factor A: Extrusion temperature (°C)	Factor B: Extrusion velocity (mm/sec)	Factor C: Filament orientation (deg.)	Factor D: Layer thickness (mm)	Response: Actual surface roughness (μm)	Response: Predicted surface roughness (μm)
1	250	250	0	0.50	2.56	2.592
2	230	50	0	0.30	3.45	3.490
3	240	150	45	0.30	2.13	2.223
4	230	50	90	0.30	2.67	2.509
5	240	150	45	0.10	2.06	2.112
6	240	150	45	0.10	2.08	2.113
7	240	350	45	0.30	2.90	2.998
8	250	50	0	0.10	3.45	3.552
9	240	150	45	0.30	3.20	3.222
10	240	150	45	0.30	2.07	2.021
11	230	250	0	0.50	2.68	2.645
12	240	150	45	0.10	2.08	2.056
13	240	150	45	0.30	2.08	2.074
14	240	50	45	0.50	3.49	3.589
15	250	50	0	0.50	3.33	3.398
16	230	250	90	0.10	2.74	2.755
17	260	150	45	0.30	2.68	2.690
18	250	250	0	0.50	2.77	2.743
19	250	50	90	0.50	2.58	2.609
20	250	50	90	0.10	2.75	2.755

(Table 4) cont.....

Trials	Factor A: Extrusion temperature (°C)	Factor B: Extrusion velocity (mm/sec)	Factor C: Filament orientation (deg.)	Factor D: Layer thickness (mm)	Response: Actual surface roughness (μm)	Response: Predicted surface roughness (μm)
21	220	150	45	0.30	2.54	2.547
22	250	250	90	0.50	2.34	2.400
23	230	50	90	0.10	3.12	3.170
24	230	250	90	0.50	3.43	3.502
25	250	250	90	0.10	3.58	3.600
26	240	150	45	0.70	2.05	2.054
27	230	250	0	0.50	3.66	3.770
28	230	50	0	0.10	3.76	3.709
29	240	150	135	0.30	3.87	3.910
30	240	150	45	0.30	2.05	2.070

Table 5. The statistical analysis of the developed model.

Statistical parameters	Sum of Squares	df	Mean square	F value	p-value Prob > F	Remarks
Model	10.151	20	0.51	24.36	<0.0001	Significant
A-Extrusion Temperature	0.0098	1	0.0098	0.47	0.5100	-
B-Extrusion velocity	1.37	1	1.37	65.64	<0.0001	-
C-Filament orientation	2.78	1	2.78	133.38	<0.0001	-
D-Layer thickness	1.18	1	1.18	56.67	<0.0001	-
AB	0.008556	1	0.008556	0.41	0.5375	-
AC	0.033	1	0.033	1.60	0.2377	-
AD	0.31	1	0.31	14.93	0.0038	-
BC	0.68	1	0.68	32.49	0.0003	-
BD	0.18	1	0.18	8.57	0.0168	-
CD	0.23	-	0.23	11.18	0.0086	-
Residual	0.19	9	0.021	-	-	-
Lack of Fit	0.18	3	0.061	94.45	<0.0001	Not significant
Pure Error	0.003886	6	0.0006476	-	-	-
Corr Total	10.33	29	-	-	-	-

The model "F-value" of 24.36 means that the developed model is statistically significant. There is only a 0.01% chance that the model "F-value" this large could occur due to noise. In addition, the value of the "p-value Prob > F" was <0.0001. The fact that the value of the "p-value Prob > F" was less than 0.050 indicates that the model terms are statistically significant. The significant model terms are B (extrusion velocity), C (filament orientation), D (layer thickness), AD (cross effect of temperature and layer thickness), BC (cross effect of extrusion velocity and filament orientation), BD (cross effect of extrusion velocity and layer thickness), as well as CD (cross effect of filament orientation and layer thickness). The "Lack of Fit" value of 94.45 implies that the lack of fit is statistically insignificant. There is only a 0.01% chance that a "Lack of Fit-F-value" this large could occur due to noise. The insignificant Fit" value implies that the model is good for a predictive purpose. From Table **6**, the values of the R-square (0.9819) and Adj. R square (0.9416) are in and all close to 1, thus, indicating that the model is suitable for correlative as well as predictive purposes.

Table 6. The Analysis of Variance (ANOVA) for the developed model.

Parameter	Value	Remarks
R-Squared	0.9819	Significant
Adj R Squared	0.9416	Significant
Adeq. Precision	15.887	Significant

The results obtained from both the numerical and physical experimentations were statistically analysed in order to obtain a predictive model, which correlates the dependent variable (surface roughness) as a function of the independent process parameters, namely extrusion temperature, extrusion speed, filament orientation, and layer thickness (Equation 1).

$$Surface\ roughness = +2.90 + 0.035A - 0.73B + 0.89C - 0.68D - 0.023AB + 0.046AC -$$
$$0.14AD + 0.21BC + 0.11BD - 0.12A^2 + 0.55B^2 + 0.55D^2 - 0.20ABD - 0.067ACD - \qquad (1)$$
$$0.097BCD + 0.65A^2B - 1.04A^2C + 0.65A^2D - 0.17AB^2$$

Where: A is the extrusion temperature (°C), B is the extrusion velocity (mm/sec), C filament orientation (deg.), and D is the layer thickness (mm).

The comparative analysis between the values of the surface roughness obtained from the physical experimentations and the ones obtained from the developed model indicates that there is significant agreement between the results obtained from the physical experimentation and the output of the model. This is shown in Fig. (**4**).

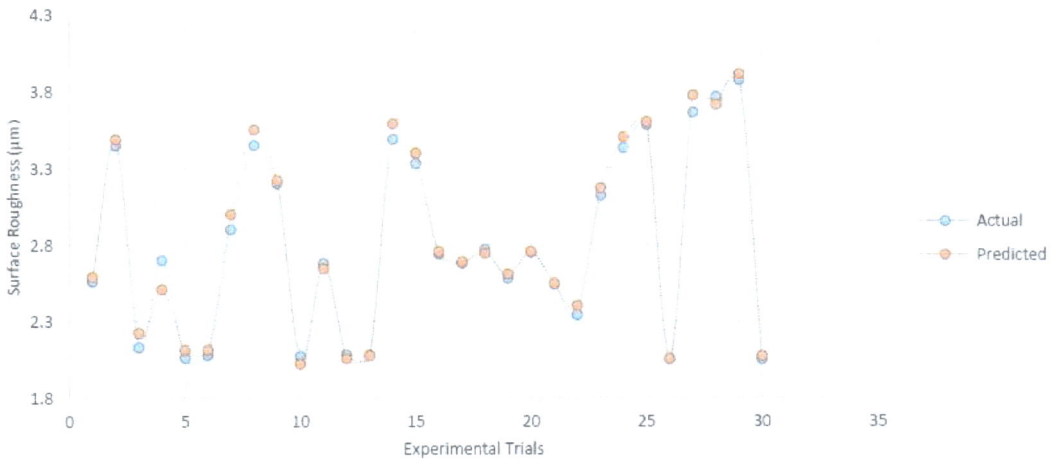

Fig. (4). The actual and predicted values of surface roughness.

Fig. (**5**) presents the error bars for the actual and predicted values of the surface roughness. The error bar depicts the graphical representation of the variations in the data obtained from the measured values of top surface roughness, which indicates the degree of uncertainty. It also indicates the preciseness of the measurement or the degree of deviation as well as the accuracy of the mean values (Cumming *et al.*, 2007). The centre of the bar is the mean of the measurement, while the top and bottom parts of the bar represent the standard deviation. The short error bar means that the standard deviation of the actual and predicted values of the surface roughness is close to the mean. This implies that there exists a minimal deviation in the magnitude of the mean from the actual and predicted values. On the other hand, the long bar indicates that there exist significant deviations in the magnitude of the mean in comparison with the actual and predicted. Fig. (**5**) indicates that the variations in the actual values and the predicted values of the surface roughness are minimal, as evidenced by the short error bars. This implies that there is a significant agreement between the actual measurements and the output of the developed model. Hence, the means of the actual and predicted values are not subjected to sampling error. In addition, the respective means are said to be representative of the magnitude of the actual and predicted values of the hardness.

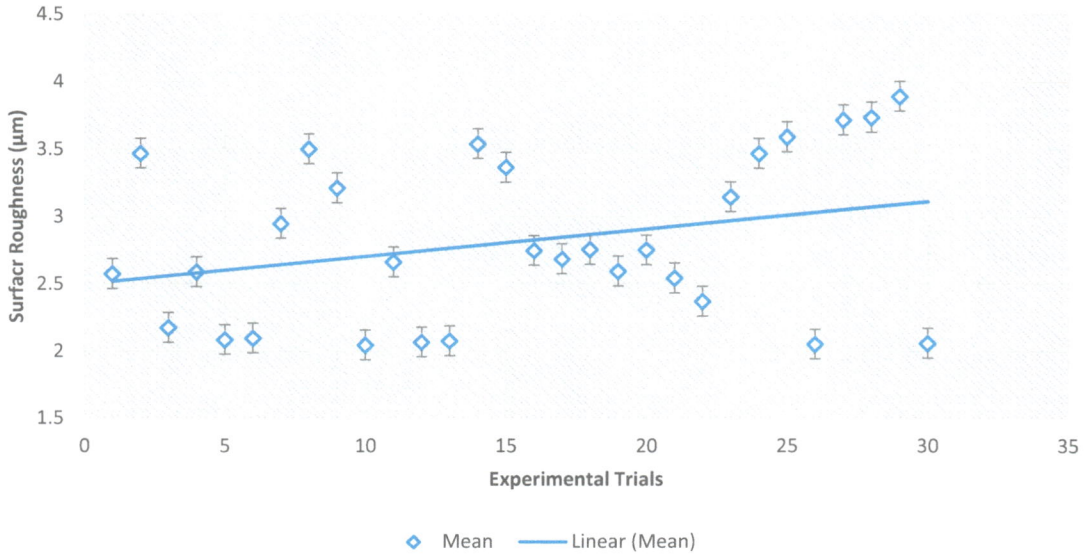

Fig. (**6**) shows the normal plot of residual for the developed model for the surface roughness. This plot indicates that there is a close relationship between the actual and the predicted values of surface roughness judging by the closeness of the data points to the diagonal line.

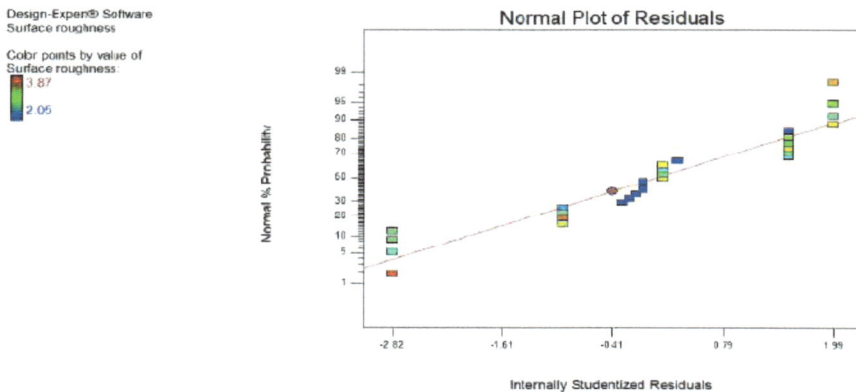

Fig. (6). The normal plots of residuals

From the physical experimentations, the combinations of the process parameters that produced the least surface roughness (2.05 µm) were: extrusion temperature (240 °C), extrusion velocity (150 mm/sec), filament orientation (45°) as well as layer thickness (0.30 mm).

Fig. (**7 - 12**) show the 3D view of the cross effect of the process parameters and the resulting surface roughness. For the cross effect of the extrusion temperature and extrusion velocity shown in Fig. (**7**), the result shows that the magnitude of the surface roughness decreases with an increase in temperature. This may be due to the fact that an increase in temperature may cause the filament to melt properly, thus resulting in lower surface roughness. Existing literature indicate that magnitude of temperature within the optimum range promotes good material flow and adhesion with a good surface finish [16]. An increase in the extrusion velocity was observed to cause the surface roughness to decrease up to the optimum value of 150 mm/sec. Beyond this point, the surface roughness was observed to increase. The higher the extrusion velocity, the higher the flow rate of the extruded material and the print speed and *vice versa*. Hence, an increase in the extrusion velocity up to the optimum value will reduce the manufacturing cycle time and ensure a good surface finish. Excessive flow of the extruded material due to an increase in the extrusion velocity may cause variation during the material deposition and binding with resulting surface irregularities. Fig. (**8**) shows the 3D plot of the cross effect of filament orientation and extrusion temperature. The results show that the two parameters, namely filament orientation and extrusion temperature are independent parameters as the variation in the magnitude of one does not affect the other. However, the filament orientation, which produced the least surface roughness, was observed at 45°.

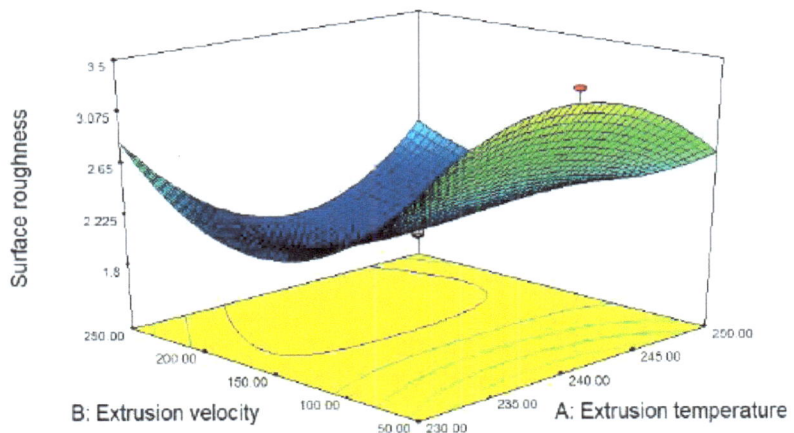

Fig. (7). The 3D plot of the cross effect of extrusion velocity and extrusion temperature,

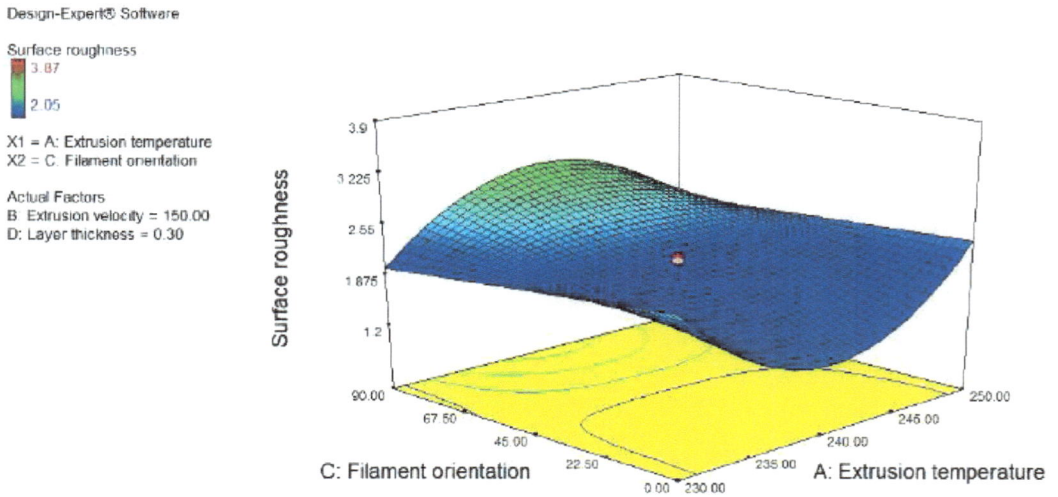

Design-Expert® Software

Surface roughness
3.87
2.05

X1 = A: Extrusion temperature
X2 = C: Filament orientation

Actual Factors
B: Extrusion velocity = 150.00
D: Layer thickness = 0.30

Fig. (8). The 3D plot of the cross effect of filament orientation and extrusion temperature.

Fig. (**9**) shows the 3D plot of the cross effect of layer thickness and extrusion temperature. The layer thickness is a measure of the thickness matrix following the addition of the filaments in layers [16]. An increase in the layer thickness beyond the optimum may promote surface roughness with the need for post processing. The higher the layer thickness, the lower the extrusion velocity, printing speed, and the overall manufacturing cycle time *and vice versa.* The 3D plot of the cross effect of filament orientation and extrusion velocity is shown in Fig. (**10**). As the extrusion velocity increases, the magnitude of the surface roughness decreases and *vice versa.* An increase in the magnitude of the filament orientation beyond the optimum value of 45° was observed to produce a significant increase in the magnitude of the surface roughness. Fig. (**11**) presents the 3D plot of the cross effect of filament orientation and extrusion velocity. An increase in the magnitude of the layer thickness and extrusion velocity brought about a significant reduction in the magnitude of the surface roughness. This implies that if the filament layers are not thick enough, during the melting process, there may be pores in the matrix, which will bring about surface roughness. Fig. (**12**) shows the 2D plot of the cross effect of layer thickness and filament orientation. Keeping other parameters constant, an increase in the magnitude of the layer thickness promotes a good surface finish.

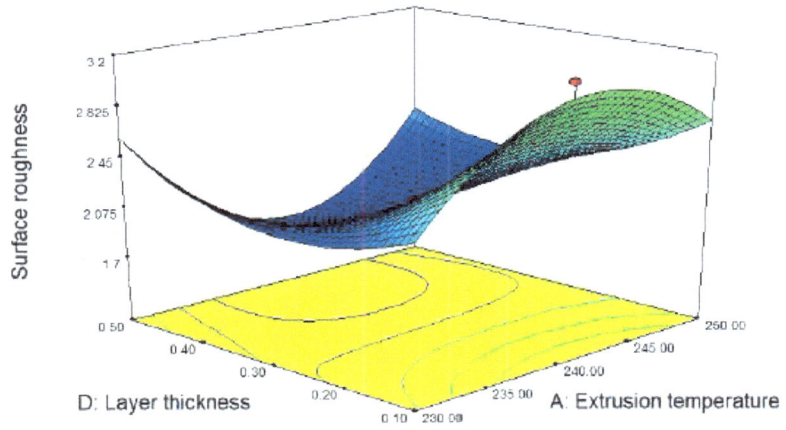

Fig. (9). The 3D plot of the cross effect of layer thickness and extrusion temperature.

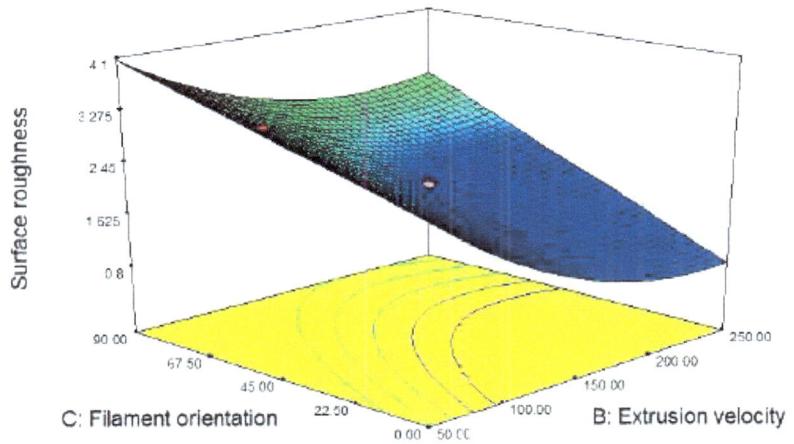

Fig. (10). The 3D plot of the cross effect of filament orientation and extrusion velocity.

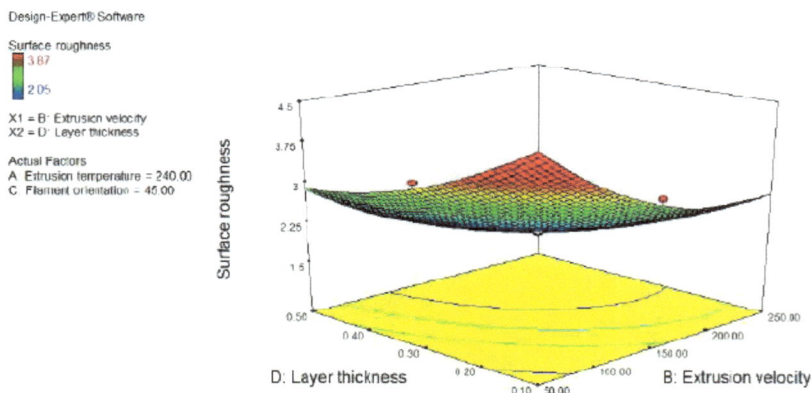

Fig. (11). The 3D plot of the cross effect of layer thickness and extrusion velocity.

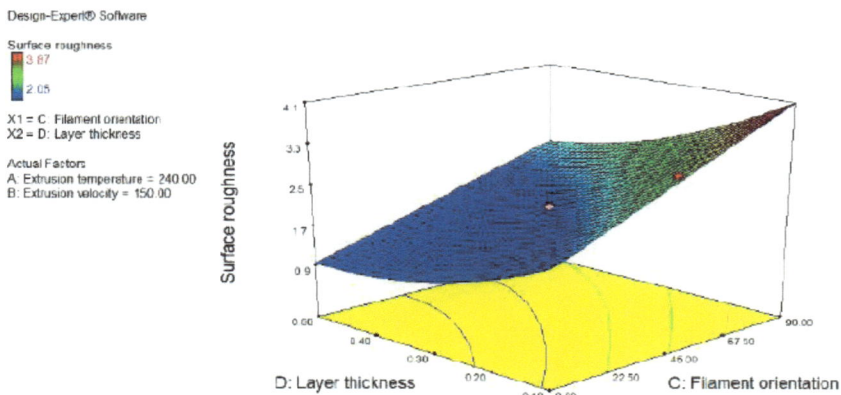

Fig. (12). The 3D plot of the cross effect of layer thickness and filament orientation.

CONCLUSION

The aim of this study was to develop the casing of a radiometer and optimise the process parameters. This was achieved using the Response Surface Methodology for the design of the experiment and FFF for the physical experimentations. The similarity in the data points of the actual and predicted values of the surface roughness from the physical and numerical experimentation respectively indicate that the developed model is suitable for the predictive purpose. The combinations of the process parameters that produced the least surface roughness (2.05μm) were extrusion temperature (240 °C), extrusion velocity (150 mm/sec), filament orientation (45°) as well as layer thickness (0.30 mm). This study provides an insight into the feasible range of process parameters that will enhance the surface finish of products developed using Polyethylene Terephthalate Glycol (PETG) filament. Future works can consider a comparative analysis of other materials that can be used for the casing development using the AM technology and the establishment of their optimum range of process parameters.

CONSENT FOR PUBLICATION

Not Applicable.

CONFLICT OF INTEREST

The authors declare no conflict of interest, financial or otherwise.

ACKNOWLEDGEMENTS

Declared none.

REFERENCES

[1] "International", In: *ASTM F2792-09e1 Standard Terminology for Additive Manufacturing Technologies* ASTM International: Pennsylvania, 2010. Available at, https://www.worldcat.org/title/standard-terminology-for-additive-manufacturing-technologies-designation-f2792-12a/oclc/900464224

[2] R. Manghnani, "The impact of additive manufacturing on the automobile industry", *Int J Curr Eng Technol.,* vol. 5, no. 3, pp. 3407-3410, 2015.

[3] J. Wang, H. Xie, Z. Weng, T. Senthil, and L. Wu, "A novel approach to improve mechanical properties of parts fabricated by fused deposition modeling", *Mater. Des.,* vol. 105, pp. 152-159, 2016.
[http://dx.doi.org/10.1016/j.matdes.2016.05.078]

[4] W. Xu, S. Tian, Q. Liu, Y. Xie, Z. Zhou, and D.T. Pham, "An improved discrete bees algorithm for correlation-aware service aggregation optimization in cloud manufacturing", *Int. J. Adv. Manuf. Technol.,* vol. 84, pp. 1-12, 2016.
[http://dx.doi.org/10.1007/s00170-015-7738-2]

[5] C. Li, S. Wang, L. Kang, L. Guo, and Y. Cao, "Trust evaluation model of cloud manufacturing service platform", *Int. J. Adv. Manuf. Technol.,* vol. 75, pp. 489-501, 2014.
[http://dx.doi.org/10.1007/s00170-014-6112-0]

[6] T.J.D. Dylan, "Design of a screw extruder for additive manufacturing", In: *A Master's Thesis submitted in partial satisfaction of the requirements for the degree in Mechanical Engineering* University of California: San Diego, 2015, pp. 1-46. Available at, https://escholarship.org/uc/item/5142579v [Accessed 4th May, 2021].

[7] S. Whyman, and K.M. Arif, "J. and Potgieter. "Design and development of an extrusion system for 3D printing biopolymer pellets", *Int. J. Adv. Manuf. Technol.,* vol. 96, pp. 3417-3428, 2018.
[http://dx.doi.org/10.1007/s00170-018-1843-y]

[8] M. Vaezi, H. Seitz, and S. Yang, "A review on 3D micro-additive manufacturing technologies", *Int. J. Adv. Manuf. Technol.,* vol. 67, pp. 1721-1754, 2013.
[http://dx.doi.org/10.1007/s00170-012-4605-2]

[9] B.N. Turner, and S.A. Gold, "A review of melt extrusion additive manufacturing processes: II. Materials, dimensional accuracy, and surface roughness", *Rapid Prototyping J.,* vol. 21, no. 3, pp. 250-261, 2015.
[http://dx.doi.org/10.1108/RPJ-02-2013-0017]

[10] A. Laplume, G.C. Anzalone, and J.M. Pearce, "Open-source, self-replicating 3-D printer factory for small business manufacturing", *Int. J. Adv. Manuf. Technol.,* vol. 85, pp. 633-642, 2016.
[http://dx.doi.org/10.1007/s00170-015-7970-9]

[11] P. Nath, J.D. Olson, S. Mahadevan, and Y-T.T. Lee, "Optimization of fused filament fabrication process parameters under uncertainty to maximize part geometry accuracy", *Addit. Manuf.,* vol. 35, no.

101331, pp. 1-12, 2020.
[PMID: 33392000]

[12]   N. Weake, M. Pant, A. Sheroan, A. Haleem, and H. Kumar, "Optimising process parameters of fused filament fabrication to achieve optimum tensile strength", *Procedia Manuf.,* vol. 51, pp. 704-709, 2020.
[http://dx.doi.org/10.1016/j.promfg.2020.10.099]

[13]   J. Chacón, M.A. Caminero, E. García-Plaza, and P.J. Núnez, "Additive manufacturing of PLA structures using fused deposition modelling: effect of process parameters on mechanical properties and their optimal selection", *Mater. Des.,* vol. 124, pp. 143-157, 2017.
[http://dx.doi.org/10.1016/j.matdes.2017.03.065]

[14]   D. Ahn, J-H. Kweon, S. Kwon, J. Song, and S. Lee, "Representation of surface roughness in fused deposition modeling", *J. Mater. Process. Technol.,* vol. 209, no. 15–16, pp. 5593-5600, 2009.
[http://dx.doi.org/10.1016/j.jmatprotec.2009.05.016]

[15]   O.A. Mohamed, S.H. Masood, and J.L. Bhowmik, "Mathematical modeling and FDM process parameters optimization using response surface methodology based on Q-optimal design", *Appl. Math. Model.,* vol. 40, pp. 10052-10073, 2016.
[http://dx.doi.org/10.1016/j.apm.2016.06.055]

[16]   I.A. Daniyan, K. Mpofu, M.O. Oyesola, and L. Daniyan, "Process optimization of additive manufacturing technology: a case evaluation of a manufactured railcar accessory", *Procedia CIRP,* vol. 95, pp. 89-96, 2020.
[http://dx.doi.org/10.1016/j.procir.2020.01.143]

[17]   A. Cattenone, S. Morganti, G. Alaimo, and F. Auricchio, "Finite element analysis of additive manufacturing based on fused deposition modeling: distortions prediction and comparison with experimental data", *J. Manuf. Sci. Eng,* 2019.
[http://dx.doi.org/10.1115/1.4041626]

[18]   M. Srivastava, and S. Rathee, "Optimization of FDM process parameters by Taguchi method for imparting for imparting customised properties to component", *Virtual Phys. Prototyp.,* vol. 13, pp. 203-210, 2018.
[http://dx.doi.org/10.1080/17452759.2018.1440722]

[19]   Y. Zhang, and K. Chou, "A parametric study of part distortions in fused deposition modeling using three-dimensional finite element analysis", *Jixie Gongcheng Xuebao,* vol. 222, pp. 959-967, 2008.

[20]   E.M. Domingo, J.M. Puigoriol-Forcada, and A.A. Garcia-Granada, "Mechanical property characterization and simulation of fused deposition modeling Polycarbonate parts", *Mater. Des.,* vol. 83, pp. 670-677, 2015.
[http://dx.doi.org/10.1016/j.matdes.2015.06.074]

[21]   Min-Ho Hong, Bong Ki Min, and Tae-Yub Kwon, "The influence of process parameters on the surface roughness of a 3D-Printed Co–Cr dental alloy produced via selective laser melting", *Appl. Sci,* 2016.
[http://dx.doi.org/10.3390/app6120401]

[22]   Y. Pupo, K.P. Monroy, and J. Ciurana, "Influence of process parameters on surface quality of CoCrMo produced by selective laser melting", *Int. J. Adv. Manuf. Technol.,* vol. 80, pp. 985-995, 2015.
[http://dx.doi.org/10.1007/s00170-015-7040-3]

[23]   I.A. Daniyan, I. Tlhabadira, S.N. Phokobye, M. Siviwe, and K. Mpofu, "Modelling and optimization of the cutting forces during Ti6Al4V milling process using the Response Surface Methodology and dynamometer", *MM Science Journal,* vol. 128, pp. 3353-3363, 2019.
[http://dx.doi.org/10.17973/MMSJ.2019_11_2019093]

[24]   I.A. Daniyan, I. Tlhabadira, K. Mpofu, and A.O. Adeodu, "Development of numerical models for the prediction of temperature and surface roughness during the machining operation of titanium alloy (Ti6Al4V)", *Acta Polytechnica Journal,* vol. 60, no. 5, pp. 369-390, 2020.
[http://dx.doi.org/10.14311/AP.2020.60.0369]

[25] I.A. Daniyan, I. Tlhabadira, S.N. Phokobye, S. Mrausi, K. Mpofu, and L. Masu, "Modelling and optimization of the cutting parameters for the milling operation of titanium alloy (Ti6Al4V)", *Proceedings of the 2020 IEEE 11th International Conference on Mechanical and Intelligent Manufacturing Technologies (ICMIMT 2020),* 2020 Added to IEEE Xplore, pp. 68-73, 2020.

[26] I.A Daniyan, K. Mpofu, and A.O Adeodu, "Optimization of welding parameters using Taguchi and Response Surface Methodology for rail car bracket assembly", *Int. J. Adv. Manuf. Technol.,* vol. 100, pp. 2221-2228, 2019.
[http://dx.doi.org/10.1007/s00170-018-2878-9]

[27] I. Tlhabadira, I.A. Daniyan, L. Masu, and K. Mpofu, "Computer aided modelling and experimental validation for effective milling operation of titanium alloy (Ti6AlV)". In: *Procedia CIRP,* 2020.

[28] B. Davoodi, and A.H. Tazehkandi, "Cutting forces and surface roughness in wet machining of Inconel alloy 738 with coated carbide tool", *Proc. Inst. Mech. Eng., B J. Eng. Manuf.,* vol. 230, no. 2, pp. 215-226, 2016.
[http://dx.doi.org/10.1177/0954405414542990]

[29] V. Aggarwal, S.S. Khangura, and R. Garg, "Parametric modeling and optimization for wire electrical discharge machining of Inconel 718 using response surface methodology", *The International Journal of Advanced Manufacturing Technology,* vol. 79, no. 1-4, pp. 31-47, 2015.
[http://dx.doi.org/10.1007/s00170-015-6797-8]

[30] M. Khajelakzay, and S.R. Bakhshi, "Optimization of spark plasma sintering parameters of Si3N4-SiC composite using response surface methodology (RSM)", *Ceram. Int.,* vol. 43, no. 9, pp. 6815-6821, 2017.
[http://dx.doi.org/10.1016/j.ceramint.2017.02.099]

[31] Wanhao, "3D Printer I3 Plus Instruction Manual", *Model: Duplicator I3,* pp. 1-60, 2020. [Online]. Available at, http://docplayer.net/53176319-3d-printer-i3-plus-instruction-manual-model-duplicator-i3-plus-visit-to-view-the-latest-version-of-this-manual.html

[32] S.A. Zmorph, "Zmorph. PETG vs PLA Filaments Comparison", https://medium.com/@ZMorph/petg-vs-pla-filaments-comparison-dfa813403ea2

[33] I.A. Daniyan, "M.O., Oyesola, K. Mpofu and A.O. Adeodu. "Thermal modelling and simulation of a screw extruder for additive manufacturing technology", *Proceedings of the 2nd African International Conference on Industrial Engineering and Operations Management Harare,* 2020, pp. 53-59 Zimbabwe

[34] I.A. Daniyan, V. Balogun, K. Mpofu, and F.T. Omigbodun, "An interactive approach towards the development of an additive manufacturing technology for railcar manufacturing", *International Journal on Interactive Design and Manufacturing,* vol. 14, pp. 651-666, 2020.
[http://dx.doi.org/10.1007/s12008-020-00659-8]

[35] I.A. Daniyan, K. Mpofu, O.L. Daniyan, F. Fameso, and M.O. Oyesola, "Computer aided simulation and performance evaluation of additive manufacturing technology for component parts manufacturing", *Int. J. Adv. Manuf. Technol.,* vol. 107, pp. 4517-4530, 2020.
[http://dx.doi.org/10.1007/s00170-020-05340-8]

[36] "Mitutuyo Quick guide to surface roughness measurement. Bulletin no 2229, 2016", [Online] Available at https://www.mitutoyo.com/wp-content/uploads/2012/11/1984_Surf_Roughness_PG.pdf,

CHAPTER 7

The Use of Sensor Based Technology for Enhancing Maintenance Operations

Ilesanmi Daniyan[1,*], **Lanre Daniyan**[2] and **Khumbulani Mpofu**[1]

[1] *Department of Industrial Engineering, Tshwane University of Technology, Pretoria, 0001, South Africa*

[2] *Department of Physics and Electronics, Adekunle Ajasin University, P. M. B. 0001, Akungba Akoko, Ondo State, Nigeria*

Abstract: The study proposes a method for integrating a sensor with a product to enhance condition based and predictive maintenance. The proposed method essentially consists of the integration of the following: sensor for measurement and data acquisition, microcontroller for control, actuator for effecting the control, cloud system for data storage, and the Internet of Things (IoT) for sharing the information collected in real time. The study presents the Proteus models for achieving measurement and control as well as the one for the integration of cloud storage and IoT. An increase in safety, high operational efficiency, and good maintenance practices can be obtained through effective analysis and use of data acquired with the use of sensors. Hence, the study presents a framework through which the use of sensors for enhancing maintenance operations can be harnessed.

Keywords: Cloud storage, IoT, Micro-controller, Proteus Model, Sensors.

INTRODUCTION

Basic sensors are used for information extraction and measurement of physical parameters from a system. On the other hand, a smart or intelligent sensor has the ability to recognise a certain condition and translate it into a quantifiable property, which is then transformed into an electrical signal for further processing [1]. The basic sensors can perform measurement of the physical objects or accept input from external sources but can neither process it nor perform any independent function, unlike smart sensors or intelligent sensors with integrated electronics. The integrated electronics in smart or intelligent sensors enhance its signal processing capabilities, logic functions, communications, detection, monitoring,

---

* **Corresponding Author Ilesanmi Daniyan:**Department of Industrial Engineering, Tshwane University of Technology, Pretoria, 0001, South Africa and Department of Physics and Electronics, Adekunle Ajasin University, P. M. B. 0001, Akungba Akoko, Ondo State, Nigeria; Tel: +27646298778; E-mail: afolabiilesanmi@yahoo.com

Ilesanmi Afolabi Daniyan (Ed.)

and effective decision-making. Smart or intelligent sensors comprise of the sensing element, signal processing unit, and microprocessor. Compared to the basic sensors, smart or intelligent sensors boast of speed, compactness, low power consumption, low maintenance cost, effective signal processing, efficiency, automation, high sensitivity and accuracy, as well as effective decision-making [2, 3].

Smart or intelligent sensors operate at high speed due to the decrease of load in the central control system. The sensor is mostly wireless, thus, making it compact with reduced weight. The presence of the digital control system makes it highly sensitive and accurate during measurement, and it is highly flexible for changes regarding the set-point and calibration from the central control computer. It can also be integrated into a network of sensors with auto-correction capability. The extracted information from smart or intelligent sensors is usually transferred to the logic system for control actions. Generally, the use of sensors finds applications in various fields such as industrial, security, automotive, rail, medical, aerospace, transport, etc. In the industrial sector, it is usually employed for measurement of the conditions of systems, system's control, as well as for diagnostic, prognostic, inspection, and operations [2]. In the security sector, smart sensors find applications in counter-terrorism such as cargo tracking and biometrics. In the aerospace, rail, and automotive sectors, it is employed for sensing and monitoring certain conditions of the system for safety and maintenance purposes. In the health sector, it is usually employed for the measurement and monitoring of the health conditions of people [2]. In the transport sector, smart sensors are employed for the control and monitoring of traffic. A few examples of sensors employed for monitoring and maintenance include vibration sensors, speed sensors, force sensors, light sensors, accelerometer, flow sensors, level sensors, current sensors, humidity sensors, temperature sensors, pressure sensors, etc. Yamasaki [4] stated that sensors promote effective communication among the physical, logical, and human worlds for effective decision-making. The physical world represents the measured or sensed system or parameter, sensor, and actuator, while the logical world represents the information processing system described by codes. The human world is the man-machine interface, which recognises the measured and the processed signals from the physical and logical worlds, respectively [4]. There exists an interrelationship between the three worlds (physical, logical, and human) for effective measurement and control. For instance, once the measurement of the system is taken in the physical world, the logical control system regulates the system *via* the actuator in the physical world, which is recognised in the human world. Hence, the interfaces between the physical and logical world are the sensors and actuators. However, recent advances in technology have witnessed a shift in the recognition roles from the human world to machine or artificial intelligence [4].

Yurish [2], in an attempt to differentiate between 'smart' and 'intelligent' sensors, explains that the smartness of a sensor relates to the technology that drives the sensor while its 'intelligence' refers to its capability or functionality. Smart sensors comprise of a combination of a sensor, an Analog to Digital Converter (ADC), an analog interface circuit, as well as a bus interface as a single unit [5, 6]. A smart sensor can be referred to as an integrated or hybrid sensor if all the components of the smart sensor are combined into one chip. The integrated sensors can perform the amplification and pre-processing of the signals from transducers before sending the signal to the microcomputers. Integrated sensors boast of several advantages such as high reliability, high compatibility, good sensitivity, and performance, as well as long term stability and cost effectiveness [7]. The essence of integrating electronics into sensors is to convert ordinary sensors to smart or intelligent sensors.

The intelligence of a sensor implies that it has one or various intelligent function (s). In other words, an intelligent sensor can independently detect and respond to a condition in the environment. Yamasaki [4] stated that the intelligent sensors have dedicated signal processing functions to enhance the flexibility of the sensing device, promote the sensing capability, reduce the load on the central processing unit, and enhance the effective distribution of information processing. Smart or integrated sensors have logic functions. It has self-testing and validating, self-adaptation and identification, self-diagnosis and compensation, as well as self-calibration capabilities [8, 9]. In other words, a sensor with integrated electronics or microprocessors is often referred to as a smart or intelligent sensor. Particularly what makes a sensor intelligent is the presence of the microprocessor or micro-controller, which interprets processes or controls the measured input signal from the sensor and the actuators that implement the control actions [10]. Intelligent sensors employ advanced signal processing techniques such as data fusion techniques, intelligent algorithms, or artificial intelligence for processing and interpretation of data. The need for intelligent sensors stems from the quest for the detection of system's abnormalities with high sensitivity and accuracy, the incorporation of predictive functions, and the sensing of physical objects from multidimensional states [4]. Other reasons include the need for a sensor with self-calibration and communication abilities, high computation ability, remote diagnosis, low energy consumption in addition to cost effectiveness, and quick response when in use [3]. Intelligent sensors thus have a good perception of the environment where it is employed and how it can be effectively managed under varying conditions.

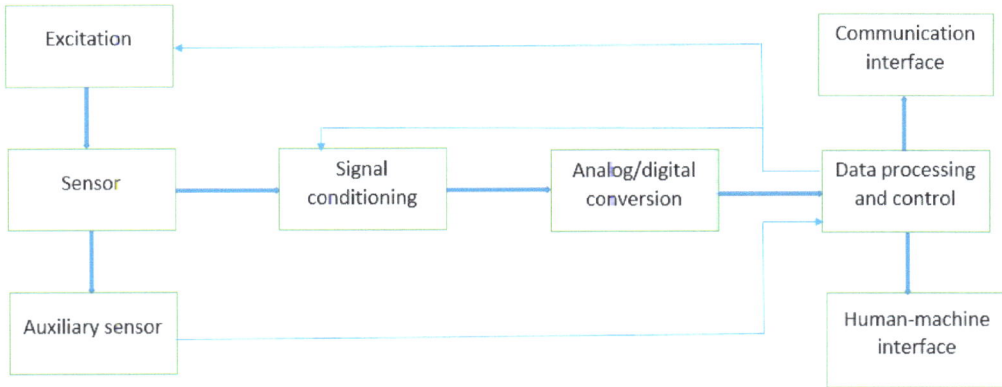

Fig. (1). The architecture of an intelligent sensor.

An intelligent sensor has conditional features, has the ability to manage its functions due to external stimuli, has the capability for advanced learning, adaptation, signal processing, analog to digital conversion in one integrated circuit, and requires specialised hardware for monitoring and control, as illustrated in Fig. (**1**). The major function of a smart or intelligent sensor includes information processing, compensation, integration, communication, validation, and data fusion. Some of the features of a smart or intelligent sensor include the presence of analog to digital converter, microcontroller with advanced features, calibrated information, data logging and real time clock, serial bus for communication, sensing or transduction element, amplifier, analog multiplexer, Analog to Digital Converter (ADC), memory and processor. However, recent advances have demonstrated the compatibility of smart or integrated sensors with cloud storage and IoT technologies for easy storage and sharing of data in real time.

Patel *et al.* [11] explain that generally, sensors can be classified into two major categories, namely passive and active sensors. For a passive sensor, no extra energy source is required, and an electric signal is produced in response to the stimulus of an external source. In other words, a passive sensor simply converts input energy to output signal energy [11]. Hence, passive sensors can only be used for energy detection when the naturally occurring energy is available. Examples of this type of sensor include infrared, thermal and electric field sensors, *etc.* On the other hand, for an active sensor, additional external sources of energy known as excitation signals are required. This type of sensor emits radiation directed towards the object of investigation and generates an output signal by adopting the necessary changes to the input signals. The sensor detects and measures the radiation reflected from that targeted object. Compared to the passive sensors, active sensors boasts of some merits, which include the ability to obtain measurements anytime, irrespective of time or season, its ability to examine wavelengths that are not sufficiently provided by the sun. However, a substantial amount of energy is required to adequately illuminate targets using the active sensors. Some examples of active sensors include laser fluorosensor and synthetic aperture radar.

Kaiser and Gebraeel [12] explained that sensor technology also finds application in prognostic health management of components or products geared towards maintenance management. This is the major focal point of this study. As one of the technologies driving the fourth industrial revolution, the use of smart sensors for measurement and data acquisition about the condition of a component is the basic building block of the Internet of Things (IoT). Sensors can be used for the development of monitoring systems to facilitate effective data acquisition for monitoring and diagnostic functions. Hence, systems are fitted with sensors for monitoring the health status of the system in real-time. Albano *et al.* [13] stated that the acquisition of complex information on the health status of a system is the enabler of advanced maintenance activities, and one of the major components required for this is the sensor. The understanding of the health status of a system in real time will enable predictive or condition based maintenance. This will enhance a decrease in the system's failure rate and its reliability in service. The data acquired with smart sensors are usually extracted, processed, and trained with specialised algorithms for effective decision-making and for future predictions. The predictive analytics with sensor technology can provide an insight into the operation and functionality of a system, thereby resulting in optimal maintenance activities coupled with improved system availability [14].

LITERATURE

The growing interest in the use of sensors technology for condition based and predictive maintenance stems from the fact that other forms of maintenance could be costly with time implications. For instance, in corrective maintenance where the systems break down before repair, there may be a payment for compensation for loss of productive time, which makes the process cost ineffective. For preventive maintenance carried out at a certain predetermined frequency, the call for maintenance may be excessive and not necessarily required. This is because preventive maintenance is usually carried out as part of the general maintenance procedure not necessarily because it is needed. Hence, preventive maintenance often triggers unnecessary costs. However, for condition-based maintenance, maintenance is carried out based on certain conditions of the systems, which act as the indicator for a call for maintenance. Predictive maintenance employs historical or real time data to predict when failure is likely to occur or when maintenance is probably required. Hence, condition based maintenance and predictive based maintenance reduce the frequency and time of maintenance and reduce system's downtime and failure, thereby increasing the system's availability, productivity, and cost effectiveness of the maintenance process. Both condition based maintenance and predictive maintenance employs sensor based technology and IoT to detect imminent machine failure, the root cause of failure before failure occurs. The diagnostic data collected in real time can be employed for troubleshooting the system in order to detect abnormalities or the root cause of failure. It can also be trained using a specialised algorithm for future predictions of failure or conditions of the system. Real time data collection *via* smart sensors and the determination of the condition of a system can influence decision making about maintenance schedules in order to maximise productivity. Palem *et al.* [15] proposed the development of condition based maintenance structure enabled by remote sensor monitoring and data capturing, real time data processing, and predictive analytics. The major component of the proposed system is the sensor array, which provides the basis for measurement and data capturing. Following data acquisition and predictive analytics, the health status of the system can be diagnosed with prescription of the possible solutions, including maintenance support. Niyonambaza *et al.* [16] proposed a predictive maintenance framework using IoT. The framework comprises four basic components, namely data acquisition with the use of sensors, data visualisation and classification, predictive analytics, as well as the selection of suitable predictive maintenance structures based on the outcome of predictive analytics. Sadiki *et al.* [17] employed the simulation approach for the comparative analyses of the performance and cost effectiveness of the traditional maintenance system with the predictive based maintenance approach. The results obtained indicated that the predictive based maintenance approach *via* the use of wireless sensor technology was more accurate and cost effective compared to the traditional maintenance approach.

Vlasov *et al.* [18] proposed a model for the optimisation of predictive maintenance of systems with the aid of wireless sensor networks. The proposed system is capable of minimising maintenance related costs with diagnostics and monitoring features. The proposed system is based on the concept of predictive maintenance with the use of sensor technology for real-time analysis of the state of a system.

With the aid of sensor technology, the development of diagnostic and prognostic models for predictive maintenance and artificial intelligence system for enhancing the maintenance of a system during its life cycle in the railcar industry have been reported [19, 20].

Many studies have reported using sensors for condition and predictive based maintenance activities [21 - 24]. However, the research on the use of sensors for maintenance-based activities is still evolving with the emerging technologies. Hence, this study proposes a method for integrating the components of smart sensors, cloud storage, and IoT for condition based and predictive maintenance. The level of integration, however, is a function of the system's modularity and analytical ability.

METHODOLOGY

The application of smart sensors for measuring the condition of a system is the building block for the internet of things and forms the bedrock for the successful implementation of a predictive maintenance framework. The basis for selecting the right sensor for measurement and data acquisition is a thorough understanding of the system's potential failure modes as well as the functional capabilities of the sensors. . It is necessary to ensure that the selected sensor has the capability to communicate effectively with other devices. It is imperative for the selected smart or intelligent sensors to share an open communication interface with other digital devices such as the cloud for interoperability and easy detection of the condition of the system. This is critical in achieving effective communication, obtaining the right information for processing, and effecting control actions. Every system has warning signals, which can be measured or monitored with the aid of sensor technology. The detection of the system's condition by sensors will either allow for effective compensation through the micro-controller and actuators or maintenance schedule [25].

Smart or intelligent sensors are selected according to their self-calibration ability, accuracy, sensitivity, self-diagnosis ability, information processing ability, fault tolerance, environment, reliability, nature of operation, and conditions to be sensed, as well as the required service life.

The proposed method of sensor's integration essentially consists of the integration of the following major components: sensor for measurement and data acquisition, microcontroller for control, actuator for effecting the control, cloud system for data storage, memory for local storage, and the IoT for sharing the information collected in real time. The IoT enables the integration of data collection and information transmission capabilities. With the IoT, the condition of the system can be monitored *via* the analysis, storage, and distribution of the data collected from the system in real time. Industry 4.0 is aimed at developing smart factories; hence the proposed sensor integration method will enable quick data acquisition, analysis, and transmission in real time. Once the measurement is taken, and data is acquired in real time with the use of the sensor, the cloud storage secures the data collected and acts as a repository for rapid access and analysis of the data captured. With the aid of cloud computing, the database provides a secure repository for rapid access and the IoT permits the sharing of the data in real time. Furthermore, the cloud storage connects to other data analytic software for the analysis of acquired data to enable quick decision-making. With the use of the Internet of Things (IoT), a vast amount of analysis can be gathered in real time, thereby leading to the development of a predictive algorithm for determining the safety, performance, or health status of the system.

The monitoring of a system in real time allows the determination of parameters whose values are outside the right values pre-set on the micro-controller. This provides decision support for condition based monitoring (CBM) and the application of artificial intelligence (whether deep or machine learning) for predictive maintenance before machine failure occurs. The system is designed to predict potential failures so that maintenance can be carried out before the sudden failure of the system.

Fig. (2) shows the application of three sensors (labelled as sensors 1, 2, and 3) for measurement and control actions. The measurement of a certain parameter represents the physical condition of the system. From the figure, the data acquired from the sensors are fed into the microcontroller (ATMEGA328P). The ULN2003 is a high power chip that contains Darlington pair for current and voltage boost. This is essential for switching systems. The system is designed to automatically actuate or effect control on the three independent systems. The normal operating condition of the system for the parameters measured is already pre-set on the micro-controller. Also, on the micro-controller, the values measured are converted to digital using the Analogue to Digital Converter (ADC). Comparative analysis is carried out vis-à-vis the pre-set normal range of the parameters. Once the measured values fall outside the range of the values, pre-set on the micro-controller, the micro-controller signals to the actuator for a chance to rectify the errors detected. Once the errors persist, there may be a need for maintenance before a catastrophic breakdown of the system.

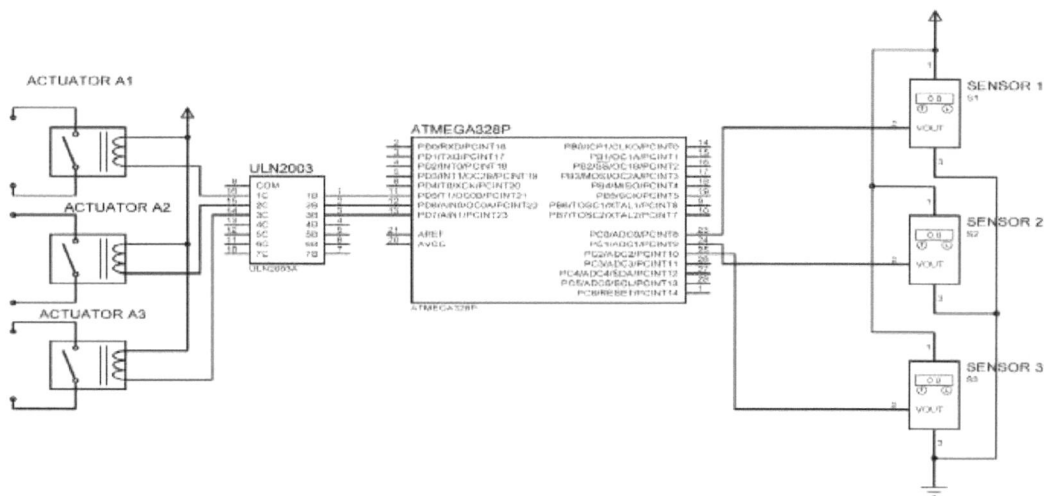

Fig. (2). The Proteus model for the integration of sensors for measurement and control actions.

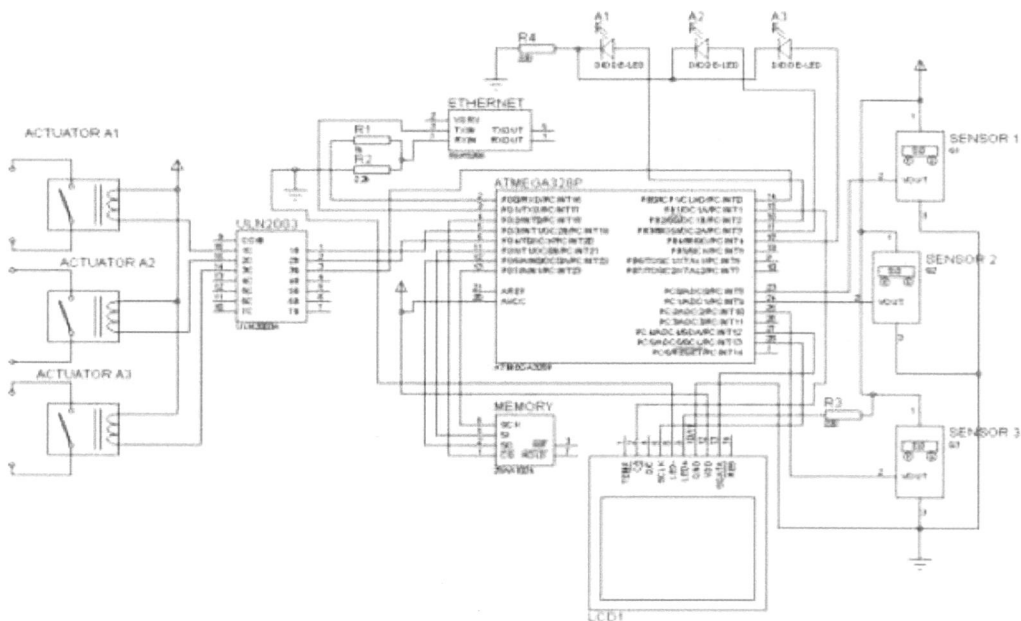

Fig. (3). The Proteus model integrating cloud storage and IoT.

Fig. (**3**) shows the integration of the measurement and control architecture with the cloud storage and the internet of things. In addition to the architecture presented in Fig. (**2**), the Proteus model in Fig. (**2**) adds the Ethernet module (ESP8266) for the internet and cloud storage, memory chips for the local storage of data, visual display for the display of the process or control parameters, storage modules, and three state Light Emitting Diodes (LEDs) to indicate the state of the actuators. The architecture is designed such that the ESP module and the codes will handle all operations related to the cloud. This is to enable condition-based maintenance as well as predictive based maintenance.

CONCLUSION

The aim of this study was to demonstrate the integration of sensor technology with other technologies such as cloud storage and IoT devices for data acquisition, processing, storage, and sharing in real time.. The study also established how condition-based and predictive maintenance could be achieved using sensor technology as a building block. To achieve this aim, the proposed method consists of the integration of major components such as sensor for measurement and data acquisition, microcontroller for control, actuator for effecting the control, cloud system for data storage, and the IoT for sharing the information collected in real time. The understanding of the health status of a system in real time will enable predictive or condition based maintenance. This will enhance a decrease in the system's failure rate and promote system's productivity, availability, and reliability in service with proper maintenance schedules. In addition, the implementation of predictive maintenance will promote cost effectiveness during maintenance operations. Future work can consider the development and performance evaluation of the proposed system.

CONSENT FOR PUBLICATION

Not Applicable.

CONFLICT OF INTEREST

The authors declare no conflict of interest, financial or otherwise.

ACKNOWLEDGEMENTS

Declared none.

REFERENCES

[1] J. Fraden, "Handbook of Modern Sensors: Physics, Designs and Applications 4th edn", Springer Science+Business Media, LLC, New York, NY 10013, USA pp. 1-631, 2010.

[2] S.Y. Yurish, "Sensors: Smart vs. intelligent", *Sensors & Transducers Journal,* vol. 114, no. 3, pp. I-VI, 2010.

[3] https://automationforum.co/what-is-a-smart-sensor-and-how-is-it-different-from-a-normal-sensor/

[4] H. Yamasaki, *What are intelligent sensors? Intelligent Sensors.* Elsevier, 1996, pp. 1-17.

[5] J.H. Huijsing, F.R. Riedijk, and G. van der Horn, "Developments in integrated smart sensors", *Sens. Actuators A Phys.,* vol. 43, pp. 276-288, 1994.

[http://dx.doi.org/10.1016/0924-4247(93)00657-P]

[6]   J.H. Huijsing, Smart Sensor Systems: Why? Where? How.*Smart Sensor Systems.,* G.C.M. Meijer, Ed., John Wiley and Sons: Chichester, UK, 2008.
[http://dx.doi.org/10.1002/9780470866931.ch1]

[7]   E.T. Powner, and F. Yalcinkaya, "From basic sensors to intelligent sensors: definitions and examples", *Sens. Rev.,* vol. 15, no. 4, pp. 19-22, 1995.
[http://dx.doi.org/10.1108/02602289510102327]

[8]   N.V. Kirianaki, S.Y. Yurish, N.O. Shpak, and V.P. Deynega, *Data Acquisition and Signal Processing for Smart Sensors* John Wiley and Sons: Chichester, UK, 2001.

[9]   R. Taymanov, and K. Sapozhnikova, "Problems of terminology in the field of measuring instruments with elements of artificial intelligence", *Sensors & Transducers,* vol. 102, no. 3, pp. 51-61, 2009.

[10]   R. Taymanov, and K. Sapozhnikova, *What makes devices and microsystems intelligent or smart* Woodhead Publisher Limited: UK, 2014, pp. 1-24.
[http://dx.doi.org/10.1533/9780857099297.1.3]

[11]   B.C. Patel, G.R. Sinha, and N. Goel, "Introduction to sensors", In: *Advances in modern sensors: Physics, design, simulation and applications.* G.R. Sinha (Edited). IOP Publishing, pp. 1-21, 2020.

[12]   K.A. Kaiser, "Predictive maintenance management using senso-based degradation models", *IEEE Transactions on Systems, Man and Cyernetics-Part A: Systems and Humans,* vol. 39, no. 4, pp. 840-849, 2009.

[13]   M. Albanao, L.L. Ferreira, G. Di Orio, P. Malo, G. Webers, and E. Jantunen, "i. Gabilondo, M. Viguera and G. Papa. "Advanced sensor-based maintenance in real-world exemplary cases", *Automatika (Zagreb),* vol. 61, no. 4, pp. 537-553, 2020.
[http://dx.doi.org/10.1080/00051144.2020.1794192]

[14]   A. Bal, and S.I. Satoglu, "Maintenance management of production systems with sensors and RFID: Case study", *Proceedings of Global Conference on Engineering and Technology Management,* pp. 82-89 Istanbul, Turkey

[15]   G. Palem, "condition-based maintenance using sensor arrays and telematics", *Internatinal Journal of Mobile Network Communications and Telematics,* vol. 3, no. 3, pp. 19-28, 2013.
[http://dx.doi.org/10.5121/ijmnct.2013.3303]

[16]   I. Niyonambaza, M. Zennaro, and A. Uwitonze, "Predictive maintenance (PdM) structure using Internet of Things (IoT) for mechanical equipment used in hospitals in Rwanda", *Future Internet,* vol. 12, no. 224, pp. 1-23, 2020.

[17]   S. Sadiki, M. Faccio, M. Ramadany, D. Amegouz, and S. Boutahari, "'Intelligent sensor impact on predictive maintenance program costs', Int", *J. Mathematics in Operational Research,* vol. 17, no. 2, pp. 170-185, 2020.
[http://dx.doi.org/10.1504/IJMOR.2020.109700]

[18]   A.I. Vlasov, P.V. Grigoriev, A.I. Krivoshein, V.A. Shakhnov, S.S Filin, and V.S. Migalin, "Smart management of technologies: Predictive maintenance of industrial equipment using wireless sensor networks", *The International Journal Entrepreneurship and Sustainability Issues,* vol. 6, no. 2, pp. 489-502, .

[19]   I.A. Daniyan, K. Mpofu, and A.O. Adeodu, "Development of a diagnostic and prognostic tool for predictive maintenance in the railcar industry", *Procedia CIRP,* vol. 90, pp. 109-114, 2020.
[http://dx.doi.org/10.1016/j.procir.2020.02.001]

[20]   I.A. Daniyan, R. Muvunzi, and K. Mpofu, "K. Artificial intelligence system for enhancing product's performance during its life cycle in a railcar industry", *Procedia CIRP,* vol. 98, pp. 482-487, 2021.
[http://dx.doi.org/10.1016/j.procir.2021.01.138]

[21]   M. Pech, J. Vrchota, and J. Bednar, "Predictive maintenance and intelligent sensors in smart dactory:

Review", *Sensors (Basel).* vol .21 , no. 1470, 1-40.

[22]   T. Brown, https://www.themanufacturer.com/articles/predictive-maintenance-sensors-and-implemen
        tation-a-solution-overview-from-dell/

[23]   S. Nirenjena, L.B. Subramanian, and M. Monisha, "Advancement in monitoring the food supply chain
        management using IOT", *Int. J. Pure Appl. Math.,* vol. 119, no. 14, pp. 1193-1196, 2018.

[24]   M. Chen, S. Mao, and Y. Liu, "Big data: A survey", *Mob. Netw. Appl.,* vol. 19, pp. 171-209, 2014.
        [http://dx.doi.org/10.1007/s11036-013-0489-0]

[25]   C. Murphy, https://www.analog.com/media/en/technical-documentation/tech-articles/choosing-the-
        most-suitable-predictive-maintenance-sensor.pdf

CHAPTER 8

Load and Temperature Significance on Tensile Strength and Flow Stress Distributions of Ecae Aluminum 6063

Temitayo Mufutau Azeez[1,2], Lateef Owolabi Mudashiru[2,*], Tesleem Babatunde Asafa[2], Adekunle Akanni Adeleke[4], Adeyinka Sikirulahi Yusuff[3] and Peter Pelumi Ikubanni[4]

[1] *Department of Mechanical and Mechatronic Engineering, Afe Babalola University, Ado-Ekiti, Nigeria*

[2] *Department of Mechanical Engineering, Ladoke Akintola University of Technology, Ogbomoso, Nigeria*

[3] *Department of Mechanical Engineering, Landmark University, Omu-Aran, Nigeria*

[4] *Department of Chemical and Petroleum Engineering, Afe Babalola University Ado Ekiti, Nigeria*

Abstract: Equal Channel Angular Extrusion (ECAE) method was considered an effective metal forming procedure to obtain higher toughness, hardness, and smooth texture. However, the magnitude of these improvements relies on extrusion load and temperature applied. This research assesses the impact of these extrusion variables on the mechanical properties and stress distributions in the Aluminum 6063 (Al 6063) produced by ECAE. Specimens of Al6063 alloy were extruded through a locally designed and manufactured ECAE die using two factors of extrusion in three levels: temperature (350 °C, 425 °C, 500 °C) and punch load (1000, 1100, and 1200 kN). The speed of the ram was held steady at 5 mm/s. The tensile strength of all extruded aluminum alloys was assessed with the universal test machine. Specimens of identical sizes and attributes were also modeled using qform software under extended applied load and temperature to investigate the distribution of stress in the extrudates. Research findings revealed that the temperature of the billet had an impact on the tensile strength more considerably than the load applied. Results of simulation revealed that more homogeneity of stress at a lower magnitude was noticed in extrudates with an increment in temperature of the billet. The simulation also reiterated the dominance of the billet temperature over the applied load on the stress dispersion with a maximum extrusion load of 500 kN at 350 °C temperature, regardless of the load applied. This result reveals how extrusion temperature increase and load enhance the tensile strength

* **Corresponding Authors Temitayo Mufutau Azeez and Lateef Owolabi Mudashiru:**Department of Mechanical and Mechatronic Engineering, Afe Babalola University, Ado-Ekiti, Nigeria and Department of Mechanical Engineering, Ladoke Akintola University of Technology, Ogbomoso, Nigeria; Tel: +2348060800687; E-mail: lomudashiru@lautech.edu.ng

Ilesanmi Afolabi Daniyan (Ed.)

of alloys but in varying degrees. An increase in load above normal level does not improve mechanical properties but is a waste of resources.

Keywords: Aluminum, Billet Temperature, Load, Simulation, Tensile strength.

INTRODUCTION

This research assesses the impact of these extrusion variables on the mechanical properties and stress distributions in the Aluminum 6063 (Al 6063) produced by ECAE. Aluminum is regarded as the second most used raw material for engineering purposes due to its special properties like strong electrical and thermal conductivity, lightweight, and corrosion resistance, among others [1]. These features are responsible for their broad range of uses in the construction, aircraft, home appliances, automotive and marine sectors. Nevertheless, aluminum components typically have poor fatigue strength and low heat resistance, partly attributable to non-uniform stress distribution within the materials [2]. Efforts have been made to enhance the mechanical and metallurgical qualities of aluminum materials. One of these methods is extrusion, which entails the displacement of metal from a die through a punch under the applied load typically supplied by a hydraulic press [3]. Extrusion involves plastic deformation and eventual structural change of aluminum into different forms based on die patterns [4]. Extruded materials are influenced by noticeable changes in mechanical and metallurgical properties during this process [5]. The Equal Channel Angular Extrusion (ECAE) approach tends to be superior to other extrusion methods [6]. This method is applicable in bulk processing aluminum products with fine grain structures. This is due to severe plastic deformation on a material under consideration and the possibility of multiple extrusion with the use of this technique, which further strengthens the established properties [7]. ECAE method involves the extrusion of material with a die of two channels and equal cross-section areas, which intersects at an angle between 90° and 150°. Deformation occurs when a piece of the metal sample (also known as a billet) is forced into the ECAE die channel under a high load [8]. The extruded material after the extrusion phase has an equal cross-section area compared to the unextruded billet [9]. This can be useful in numerous industrial applications that require fine grain texture in larger quantities [10]

Numerous studies have reported that the ultra-fine grain products produced by ECAE are capable of retaining a blend of adequate ductility and high strength [11]. These excellent mechanical products are highly desirable over the next decade in the manufacture of advanced structural materials [12]. However, the achievement of these properties results from further refinement to create a precise microstructure sufficient to improve the ductility of the substance [13]. ECAE components are characterized by non-uniform strain distribution and a high load of extrusion requirement during the process [14]. These challenges emerge due to an uneven spread of stress in extruded products, resulting in low production efficiency, and consequently, a rise in total costs. These challenges are exacerbated by an insufficient or excess load of the extrusion and the temperature of the billet. If the load of the extrusion and the temperature of the billet are higher than appropriate, the component under consideration may be damaged easily, with more manufacturing time required and the cost of processing increased [15]. Inadequate load and temperature often decline the physical property of the material, like the tensile strength [16], and consequently lead to uneven distribution of flow stress in the extruded product. The stress flow of material is also of special significance for the extrusion of complicated sections. The concept of material flow stress during the extrusion process was of considerable concern to aluminum alloys in particular [16]. Thus, it is important to examine the thermal and load response of extruded aluminum to flow stress using the ECAE process.

MATERIALS AND METHOD

Experimental Procedure

The method involves measurement of the tensile strength of Aluminum 6063 after extrusion, varying extrusion factors such as load and temperature, followed by a process simulation to assess the effect of these two parameters on the stress distribution of extruded aluminum 6063. Specimen of aluminum 6063 was machined into billet (11.95x 11.95 x 40 mm^3) and then milled to achieve a smooth surface finish. The chemical composition of the material was measured with the use of a Glow Discharge Mass Spectrometer (PEM 2380). The sample size was selected to conform with the configuration of the equal channel angular extrusion die of 45 and 12 x 12 mm^3 (length x breadth x height) cross section with a die angle of 120°, as presented in Fig. (1). This provides adequate entry clearance for the aluminum billet. The tool steel was machined into a die (Fig. **2 a and b**) and punched (Fig. **3 a and b**) using lathe and drilling machines.

The billet samples were washed, placed into the electric furnace, and heated to the specified range of temperatures of 350-500°C at three pre-extrusion treatment stages. The die wall was lubricated with Jatropha oil for thermal insulation and increased the flow of material in the die. The heated aluminum workpiece was then placed into a lubricated die with a punch on it to push the billet out from the die using a hydraulic press. The entire assembly of the billet, punch, and die was mounted in the hydraulic press for extrusion under variable loads (1000 to 1200 kN at three treatment levels). For each pass, the ram speed of 5 mm/s was set. A universal testing machine (UTM) was used to conduct a tensile test on extruded aluminum 6063 samples that were already machined to an appropriate size. The pulling force of 25kN was applied to the aluminum axially. Each test was conducted three times to assure the consistency of the result.

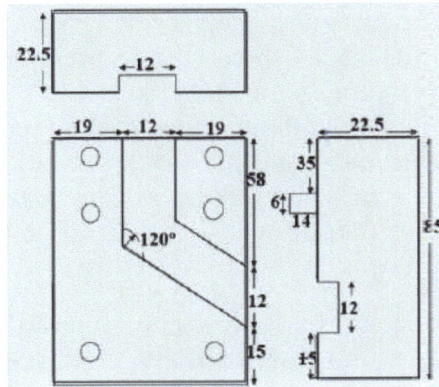

Fig. (1). Die configuration.

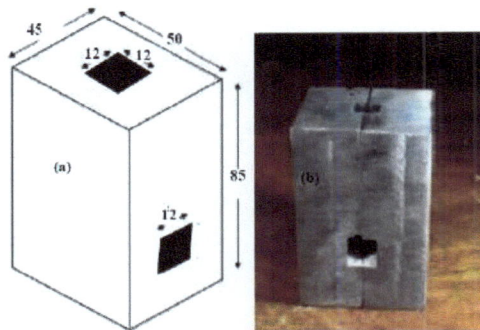

Fig. (2 (a & b)). Die (a): design in closed position (b) fabricated.

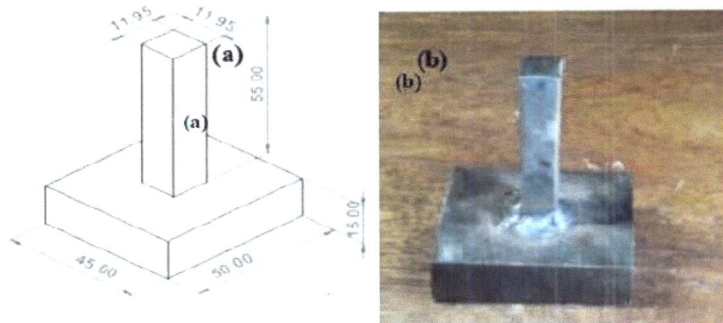

Fig. (3 (a &b)). Punch (a): design (b) fabricated.

Finite Element Simulation

The ECAE process simulation was carried out on the qform platform-professional finite element-based program devoted to the simulation, analysis, and optimisation of metal forming procedures. During the simulation, the stress distribution within the extrudates was analysed at varying temperatures and loads. The simulation process consists of three stages: preprocessing, processing, and post-processing. At the pre-processing stage, structures similar to those used in the experiments were generated, although few assumptions were applied, such as the homogeneous and isotropic visco-plastic deformation. The hydraulic press supplied compressive force at an extrusion rate of 5 mm/s, while the friction between the workpiece and the die was assumed to be insignificant due to the application of lubricant. The cooling medium was air.

The overall material flow model for extrusion in *qform* platform comprises of the dynamic equation (Eq. 1), the compatibility condition (Eq. 2), constitutive equation (Eq. 3), incompressible equation (Eq. 4), energy balance equation (Eq. 5), and flow stress (Eq. 6) [17].

$$\sigma_{ij,j} = 0 \tag{1}$$

$$\varepsilon_{ij} = \frac{1}{2} \left(V_{i,j} + V_{j,i} \right) \tag{2}$$

$$\sigma_{ij} = \frac{2}{3} \frac{\bar{\sigma}}{\bar{\varepsilon}} \varepsilon_{i,j} \tag{3}$$

$$V_{i,i} = 0 \tag{4}$$

$$\rho c T = \left(k\, T_i \right), i + \beta\, \bar{\sigma}\dot{\bar{\varepsilon}} \tag{5}$$

$$\bar{\sigma} = \bar{\sigma} \left(\bar{\varepsilon}, \dot{\bar{\varepsilon}}, T \right) \tag{6}$$

Where: σ_{ij}, ε_{ij}, $V_{i,j}$ are respectively stress, strain rate, and velocity components.

$\bar{\sigma}, \bar{\varepsilon}, \dot{\bar{\varepsilon}}$ are respectively effective stress, effective strain, and effective strain rate.

ρ, c, k are respectively density, specific heat, and thermal conductivity.

T is the temperature, and β is the heat generation efficiency typically estimated to be 0.9 - 0.95

All the input variables employed for the simulation were derived from the ECAE tests carried out earlier [17]. The angle of die employed for simulation was 120°

while the billet was loaded with 1000, 1100, and 1200 kN at 350, 425, and 500 °C, respectively.

RESULTS AND DISCUSSIONS

Result of Tensile Strength

Table 1 displays the tensile strength measures of the extruded specimens based on 9 experimental runs. The tensile strength effects were measured when loads of 1000 kN, 1100 kN, and 1200 kN were kept steady, and temperatures of 350 °C, 425 °C, and 500 °C were varied. The tensile strength was within the range of 246-303 MPa. The parent material had a tensile strength value of 243 MPa [18]. There was a huge difference in the tensile strength of the extruded product when the temperature was varied. The tensile strength response to the load effect wasnot too pronounced. For example, at a load of 1000 kN but changing the extrusion temperature from 350°C to 500°C, the tensile strength was increased from 250 MPa to 303 MPa. This may be attributed to recrystallisation and grain growth at higher temperatures [19]. Whereas at a steady temperature of 350°C but raising the load from 1000 kN to 1200 kN, the tensile strength was increased from 250 MPa to 251 MPa.

Table 1. Tensile values of extruded aluminum alloy.

S/N	Load (kN) Temp (°C)		TS (MPa)
1	1000	350	250
2	1000	425	298
3	1000	500	303
4	1100	350	246
5	1100	425	286
6	1100	500	302
7	1200	350	251
8	1200	425	295
9	1200	500	303

The measurement of tensile properties was carried out on non-extruded samples (source material) and the extruded samples using an equal channel angular extrusion technique to evaluate the stress-strain behavior of the Al 6063. The findings demonstrated similar stress strain behaviour in all samples at selected temperatures, and those behaviours are shown in Fig. (4). This illustrated the real stress-stress diagram of the Al6063 alloy. A huge increase in stress wasobserved at 350 °C. The relative increase in stresses was noted at 425 and 500 °C.

Characteristics of the plot suggested that the yield strength and ultimate tensile strength (UTS) increased after the third ECAE passed. Necking started shortly after yielding. This is in strong alignment with the findings mentioned earlier [20].

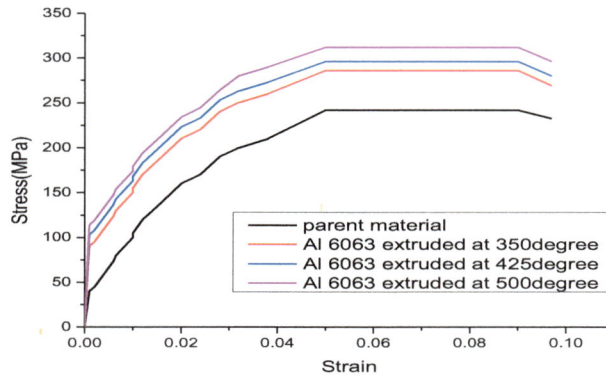

Fig. (4). Tensile behavior of the extruded samples and parent material.

Temperature Effects on the Flow Stress Distribution

Fig. (**5a-5d**) shows the distribution of stress in billets after extrusion at specified temperatures of 350,425, 500, and 550°C and a steady load of 1000 kN. Maximum stress is reduced with an increased extrusion temperature, which means that the temperature decreases the stress level. The metal yield strength is often decreased with high temperatures due to increased thermal stress. At this temperature, the metal has to go through microscopic plastic deformation, thus losing a portion of the residual stress [21].

The higher flow stress at 350°C relative to the stress before extrusion is caused by the plastic deformation the billet was subjected to. Except for the top edges of the extrudate, where maximal stress of 172 MPa has been observed, the flow stress is relatively homogenous between 130 and 135 MPa (Fig. **5a**). At a higher extrusion temperature of 425°C, the maximum and minimum stress values respectively decreased to 167.8 and 128 MPa (Fig. **5b**). Similar occurrence for extrusion was observed at 500°C with a maximum and minimum stress level of 142.4 and 103 MPa, respectively (Fig. **5c**). At 550 °C, the stress level declined significantly to 59 and 104 MPa, respectively, for the minimum and maximum levels (Fig. **5d**). A measured increase in temperature with a decrease in stress level was observed in Fig. (**6a**). A massive decrease in stress level as the temperature of the extrusion reaches 550 °C just 50 °C before the melting point of the aluminum (600 °C) was observed. The plot also refers to the greater influence of temperature as a key criterion in the ECAE system, particularly for aluminum alloys [22].

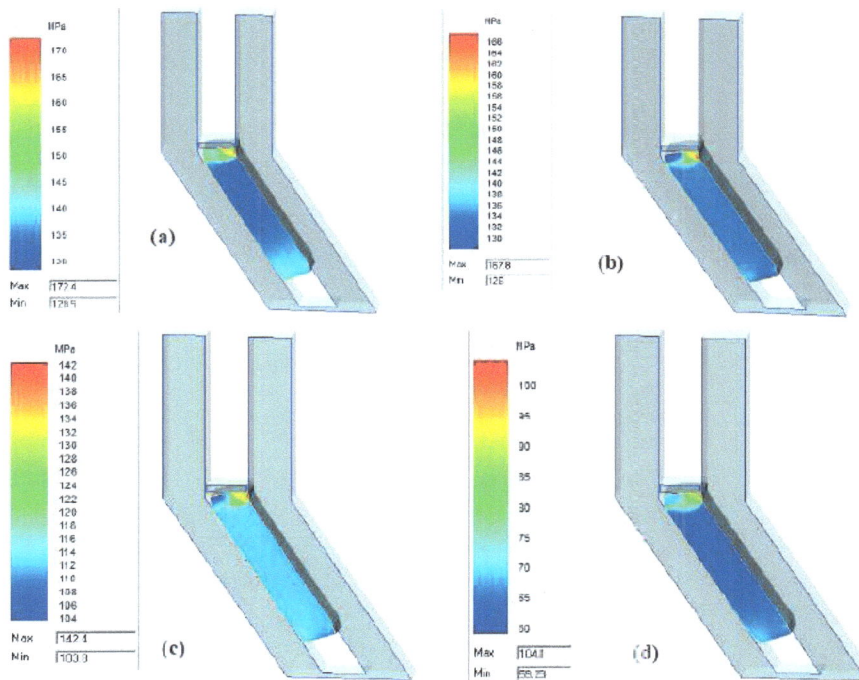

Fig. (5 (a-d)). Flow stress distribution in samples extruded at temperature of (**a**) 350 °C (**b**) 425 °C (**c**) 500 °C and (**d**) 550 °C.

The stress distribution tends to be more uniform with an increase in extrusion temperature, as seen in Fig. (**5a-d**). Appropriately, the upper sections of the billets should be trimmed off to provide a product of more uniform stress distributions. Such an insignificant stress gradient creates high-strength aluminum with higher tensile strength, as seen in the experimental results in Table **1**, where the tensile strength rises with the extrusion temperature [23].

Fig. (**6b**) displays the actual extrusion load *versus* the displacement of billets extruded at various temperatures. The response to the deformation has three distinct phases [24]. At first, the load of the extrusion rises exponentially, signaling the downward displacement of the billets to the corner of the die. This is instantly followed by the dwelling of the extrusion load, referring to the steady-state of the system, as the billets flow through the angular section of the die continuously. This load gives the maximum load of the extrusion at that specific temperature.

Eventually, the load of the extrusion declines to zero, signaling the end of the deformation process. As noted, the extrusion load declines at a rising extrusion temperature [17], with a peak load of 500 kN required for extrusion at a temperature below 500°C. At 500°C and above, the load of the extrusion decreases significantly. For instance, while the extrusion load reduced to 430 kN at 500°C, the load dropped below 260 kN at 550°C . This substantial drop in the extrusion load can be based on the fact that the flow stress declines with a rising temperature [25] and, by implication, the processing time also decreases [26].

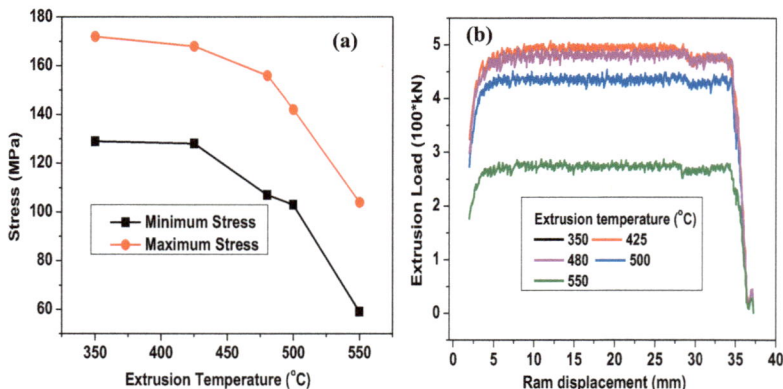

Fig. (6 (a &b)). (a) Stress against extrusion temperature (b) extrusion load – ram displacement curves for extrusion carried out under an applied load of 1000 kN at different temperatures. Curves corresponding to 350°C and 425°C overlapped.

Loading Effects on Flow Stress Distribution

From the simulations carried out, the application of a load greater than 500 kN does not impact the extrusion load of the ECAE system, while the distribution of stress is marginally influenced. This is verified in Fig. (**7a-b**), where the stress difference for extrusion at a constant temperature of 350 °C but a varying load of 700 kN and 1000 kN is just 12.4 MPa with higher stress at 700 kN. Also, the distribution of stress tends to be more uniformly distributed at 700 kN relative to 1000 kN. This is possibly due to the lower deformation rate and the consequent lower strain gradient [27]. Fig. (**7 c-d**) also affirms that the extrusion loads are equal considering the big variation in the applied load, provided that the extrusion temperature remains stable. At 350°C, the applied loads of 700, 1000, 1100, and 1200 kN possess a similar extrusion load of 500 kN. However, the same selection of applied load at 500 °C has an extrusion load of 4300 kN. Theoretically, the maximal extrusion load for an aluminum billet deformed at a recrystallization temperature of 350°C is 508 kN [28]. This is also verified by the simulation performance.

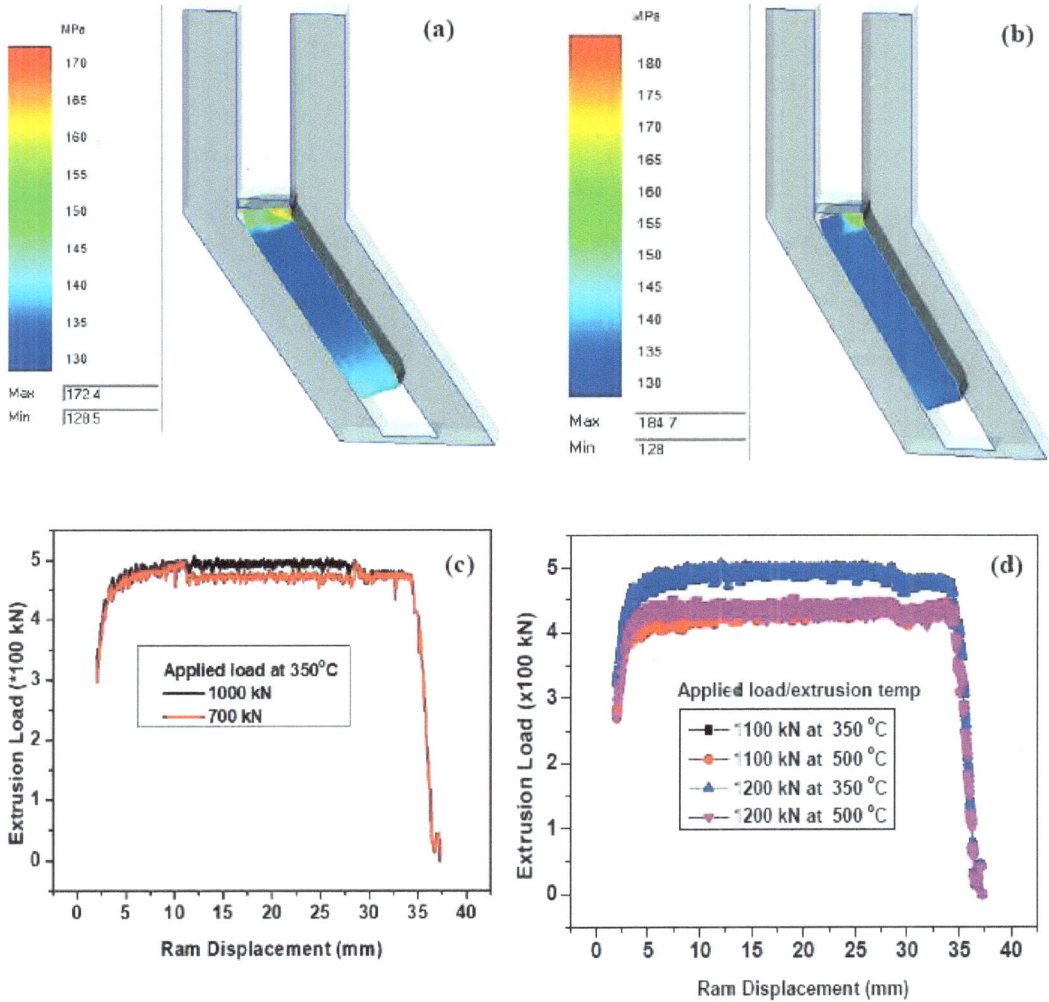

Fig. ((7a-d)). Stress distribution in samples extruded at 350°C under an applied load of (a) 1000 kN (b) 700 kN. (c) The extrusion load-displacement curve and extrusion carried out under an applied load of 1100/1200 kN at two temperature regimes are shown in (d) Although higher loads can minimise the extrusion time, they do not influence the actual extrusion load, which hugely dictates the power absorbed during the metalworking process [26].

CONCLUSION

The following hypotheses can be reached based on the different outcomes obtained and the findings of this report.

i. The temperature has a substantial influence on rising the tensile strength of the extruded aluminum alloy.

ii. Aluminum treated at 500 °C, 1000 kN offered the extruded sample a homogeneously improved thermal stress spread, and the stress level is very low relative to the other two temperatures of 425 °C and 350 °C at the equal load.

iii. The temperature has load reduction potential at a temperature of 500 °C and above but less than the melting temperature.

CONSENT FOR PUBLICATION

Not applicable.

CONFLICT OF INTEREST

The authors declare no conflict of interest, financial or otherwise.

ACKNOWLEDGEMENTS

Declared none.

REFERENCES

[1]   D.E. Esezobor, and S.O. Adeosun, "Improvement on the strength of 6063 aluminum alloy by means of solution heat treatment", *Mater. Process. Challenges Aerosp. Ind,* no. 1, pp. 645-655, 2006.

[2]   M.A. Nurul, and S. Syahrullahi, "Study of alternative lubricant for cold extrusion process of A1100 pure aluminum", *J. Teknol.,* vol. 71, no. 2, pp. 139-143, 2014.

[3]   J.K. Weertzman, "Hall-petch strengthening in nanocrystalline metals", *Mater. Sci. Eng.,* vol. 1, no. 2, pp. 161-167, 2012.

[4]   R. Mohan, M.I. Santhosh, and K.G. Venkata, "Improving mechanical properties of al 7075 alloy by equal channel angular extrusion process", *Int. J. Mod. Eng. Res.,* vol. 3, no. 5, pp. 2713-2716, 2013.

[5]   R.S. Rusz, and K. Malanik, "Using severe plastic deformation to prepare ultra-fine grained material by ECAE method", *J. Achiev. Mater. Sci. Eng,* no. 28, pp. 683-687, 2007.

[6]   A.A. Abiodun, S.I. Ojo, and I.D. Abimbola, "Equal channel angular extrusion charateristics on mechanical behavior of aluminum alloy", *Intech Opens,* vol. 4, pp. 264-277, 2017.

[7]   D.V. Murty, and M. Ramulu, "Deformation study of dual equal channel lateral extrusion", *Int. J. Eng. Stud,* no. 3, pp. 161-168, 2009.

[8]   S. Raghavan, H. Qingyou, S. David, and G. Percy, "Continuous severe plastic deformation processing of aluminum alloys", [Online]. Available: www.osti.gov,

[9]   R.A. Parshikov, A.I. Rudskoy, A.M. Zolotov, and O.V. Tolochko, "Technology problem of equal channel angular pressing", *J. Adv. Mater. Sci,* no. 24, pp. 26-36, 2013.

[10]   A. Roschowski, "Processing metals by severe plastic deformation", *Diffus. Defect Data Solid State Data Pt. B Solid State Phenom.,* no. 10, pp. 13-22, 2005. [http://dx.doi.org/10.4028/www.scientific.net/SSP.101-102.13]

[11]   M.L. Alexander, V. Alexander, and Y.V. Pandolena, "Analysis of equal channel angular extrusion by upper bound method and rigid block model", *J. Mater. Res.,* vol. 7, pp. 359-366, 2014.

[12]   K. Padmanathan, *Parameter optimization of the process of AA6xxx and AA7xxx series aluminum extrusion.* Auckland University of Technology, 2013.

[13] S. Xu, "Finite element analysis and optimization of equal channel angular pressing for producing ultra-fine grained materials", *J. Mater. Process. Technol.,* no. 184, pp. 209-216, 2006.

[14] O. Ahmed, "Effect of parameters and operating conditions on flow of extrusion load", *Scr. Mater.,* no. 49, pp. 46-55, 2008.

[15] B. Avitzur, *Metal forming processes and analysis.* 2nd ed. Wiley interscience: New york, 2007.

[16] C.B. Seung, E.H. Yuri, S.K. Hyoung, T.J. Hyo, and J.H. Raph, "Calculation of deformation behavior and texture evolution during equal channel angular pressing", *Mater. Sci. Forum,* vol. 408, pp. 697-702, 2002.

[17] B. Nickolay, V. Sergey, and V. Alexey, *Material forming simulation environment based on QFORM software system* Moskow, 2014.

[18] I.U. Mohammed, and K.S. Senthil, "Application of response surface methodology in optimizing process parameters of twist extrusion process for aluminum aa 6061- T6 alloy", *Measurement,* no. 94, pp. 126-138, 2016.
[http://dx.doi.org/10.1016/j.measurement.2016.07.085]

[19] H. Bakhtiari, K. Mahdi, and R. Sina, "Modeling analysis and multi-objective optimization of twist extrusion process using predictive models and meta-heuristic approaches, based on finite element results", *J. Intell. Manuf.,* vol. 10, no. 2, pp. 248-256, 2014.

[20] S.A. Akbari, S.H. Mousavi, and A. Bahadori, "Numerical and experimental studies of the plastic strains distribution using subsequent direct extrusion after three twist extrusion passes", *Mater. Sci. Eng.,* vol. 527, pp. 3967-3974, 2010.
[http://dx.doi.org/10.1016/j.msea.2010.02.077]

[21] R.A. Winholtz, "Residual stresses: macro and micro stresses", *Mater. Sci. Technol.,* no. 49, pp. 46-55, 2001.

[22] L. Zhenhua, C. Xianhua, and S. Qiangian, "Effects of heat treatment and ECAE process on transformation behaviour of TiNi shape memory alloy", *Mater. Lett.,* no. 59, pp. 705-709, 2004.

[23] Z. Maoyu, M. Zhengzheng, T. Chunyan, and L. Ping, "The relationship between tensile strain and residual stress of high strength dual phase steel sheet", *MATEC Web of Conferences,* vol. 175, no. 01033, pp. 1-5, 2018 .

[24] J. Nemati, S. Sulaiman, G.H. Majzoobi, B.T.H.T. Baharudin, and M.A. Azmah, "Finite element study of deformation behavior of Al-6063 alloy developed by equal channel angular extrusion", *Adv. Mat. Res.,* no. 1043, pp. 119-123, 2014.

[25] I. Flitta, and T. Sheppard, "Effects of pressure and temperature variations on FEM prediction of deformation during extrusion", *Mater. Sci. Technol.,* vol. 21, no. 3, pp. 339-346, 2005.
[http://dx.doi.org/10.1179/174328405X29221]

[26] F.A. Lontos, D.A. Soukatzidis, D.A. Demosthenous, and A.K. Baldoukas, Effects of extrusion parameters and die geometry on the produced billet quality using finite element method*international Conference of Manufacturing Engineering*, 2008, pp. 215-228.

[27] J. Jandrlic, S. Reskovic, T. Brlic, and V. Furlan, Effect of deformation rate on low carbon steels mechanical properties*in Materials and Engineering*, 2018, pp. 1-6.
[http://dx.doi.org/10.1088/1757-899X/461/1/012030]

[28] X. Wang, M. Zhang, N. Tang, N. Li, L. Liu, and J. Li, A forming load prediction model in BMG micro backward extrusion process considering size effect.*in Physics of Non Crystalline Solid*, 2013, pp. 146-151.

CHAPTER 9

Development of Lean Assessment Tools for Maturity Evaluation in Warehouse Environment

Adefemi Omowole Adeodu[1,*], Mukondeleli Grace Kanakana-Katumba[1] and **Rendani Maladzhi[1]**

[1] Department of Mechanical and Industrial Engineering, University of South Africa, South Africa

Abstract: The aim of this research is to develop lean assessment tools and the corresponding practices for maturity evaluation within a warehouse environment operating mainly as third party logistics (3PL). The objective is to examine the performances of the warehouses in terms of productivity, quality, and employee satisfaction. The research is a case study in nature, where design-thinking approach was applied for the development of lean tools. The study was conducted *via* literature reviews and multiple case study assessments of the implementation of lean practices in third party logistics (3PL) services. Thus, both qualitative and quantitative approaches were used. Furthermore, there was a combination of exploratory and descriptive characters in the study. Lean assessment tools that can be used in warehouse environments were developed. From the lean maturity evaluation of the warehouses, overall maturity is averagely implemented. Analyses of the warehouse productivity show good implementation with a score of ≥80-90%. There is a direct relationship between lean practices and productivity. Warehouse quality falls within the average level of implementation with a score of ≥60-80%. Employee satisfaction has little or no effect on overall maturity with the score within the average level of implementation of ≥60-80%. The study developed tools for lean maturity assessment adapted to third party logistics and the total supply chain warehouse environment *via* design thinking. Consequently, the research outcomes were two-fold, filling the gap in the development of comprehensive warehouse lean maturity assessment tools and providing insight into the actual implementation of lean warehousing.

Keywords: Lean, Manufacturing, Maturity, Third party logistics, Warehousing.

INTRODUCTION

Over two and half decades, many organisations, mostly in the developed nations, have focused on the adoption of various organisation improvement paradigms of cost reduction and profitability increase [1,3]. Total Quality Management (TQM), Six Sigma, Lean manufacturing, and re-engineering are just a few of the improve-

* **Corresponding Author Adefemi Omowole Adeodu:** Department of Mechanical and Industrial Engineering, University of South Africa, South Africa; Tel:+27787223718; E-mail: eadeodao@unisa.ac.za

Ilesanmi Afolabi Daniyan (Ed.)

ment strategies that have emerged throughout the years and are now widely used by organizations [1, 3 - 4].These strategies focus on various aspects of quality improvement [3], elimination of variation, waste reduction, organisational restructuring, problem solving, cost cutting, *etc.* [1].

Lean manufacturing is a multidimensional production optimisation approach that captures various management practices aimed at waste reduction and improving operational effectiveness [5]. The evidence of adoption of the approach in various sectors of the economy is not limited to quality and productivity improvement but also considers factors such as supportive learning environment and leadership development in organisation [3]. Although lean was developed originally to minimise waste in a manufacturing environment, but recently, firms with diverse professional areas are increasingly adopting lean techniques, for instance, in service and logistics industries [6 - 9]. Lean has also crept into warehousing [1, 2] [10 - 13]. Companies like Philips, Bosch, Menlo, and CEVA logistics earlier adopted the programme in order to encourage warehouses to adopt lean management practices and techniques [2, 6, 8]. Most industries that have adopted this idea and invested significant resources, especially logistics, face the challenge of making it effective in terms of maturity.Several tools have been developed by practitioners and researchers to measure lean maturity in manufacturing environments [14 - 16], but there is a dearth of assessment tools and literature when it comes to lean maturity in the warehousing environment. According to Visser [2], a well-refined model to measure lean maturity in the production environment was developed by Philips, which was later engrafted into Philips warehousing. The shortcoming of the model was that it fitted more to the production environment than warehousing. The major objective of lean warehousing is to improve the delivery of added value towards customers, suppliers, and internal performance processes [1, 2].

Technical literature on warehouse lean maturity assessment is meager [11]. Three categories can be identified: articles that propose a framework for designing or analysing warehouses, those that directly address performance, and researches conducted on the development of assessment tools for lean maturity measurement. Rouwenhorst *et al.* [17] and Gu *et al.* [18] belong to the first category. Co-ordination problems due to the investigation of warehousing subproblems were addressed. According to Rouwenhorst *et al.* [17], a framework to accommodate these problems was suggested, but it was largely descriptive in nature and did not cater to an operational technique to coordinate the design decision. As for Gu *et al.* [18], the decision problems associated with design and operation were categorized rather than the overall warehouse performance assessment. The second category was based on logistics benchmarking with few warehouse benchmarking results. Exceptional results were accredited to Stank *et al.* [19],

Cohen *et al.* [20], and Hackman *et al.* [21]. According to Stank *et al.* [19], survey data were gathered from 154 warehouses to determine their implementation of benchmarking and the area of implementation. The authors considered the sizes of the companies and services provided to determine the correlations with benchmarking practices and operations. Contrary to the work of Cohen *et al.* [20], whereby varieties of performance metrics were used to evaluate service parts warehouses. The deficiency in the work wasthe ambiguity in the results that described the relationship between performance and inventory practices. Hackman *et al.* [21] developed a model of a warehouse environment in which labour, space, and investment were used as resources, broken case lines, full case lines, pallets lines, accumulation, and storage functions served as services produced. The model was based on the answers to the three main questions: do larger warehouses perform more efficiently? Do capital-intensive warehouses perform more efficiently? Do non-union facilities outshines their union counterparts? Using Data Envelopment Analysis (DEA) model to quantify efficiency based on data gathered from 57 warehouses in operation between 1992 and 1996, Hackman *et al.* [21] concluded that smaller, less capital-intensive warehouses are more efficient and that unionization has less impact on efficiency. The third category was a more recent work carried out by Collins *et al.* [22], which described the collection of warehouse metrics, *e.g.,* picking, inventory accuracy, storage speed, and order cycle time, used in a multi attribute utility theory analysis. De koster & Watfenious [23] also performed an international comparison across a set of 65 warehouses in three continents in 2000 that used various performance measures to identify differences in performances of warehouses operated by third party logistics providers or self-operated warehouses. The authors' conclusion was that there are similarities in performances across countries and operating parties. De Koster and Balk [24] carried forward the research of Koster and Warffemius [23] by gathering data on 39 of the 65 previously analysed warehouses using DEA. The inference from the work was that European warehouses that ran with 3PL techniques were more efficient than Asian and American warehouses.

The aim of the research is to develop lean assessment tools and the corresponding practices for lean maturity evaluation within a warehouse environment operating mainly as third party logistics (3PL). The objective is to examine the performances of the warehouses in terms of productivity, quality, and employee satisfaction exploring the concept of design thinking to influence the lean warehousing design and analyse the design process. Design thinking is a way of thinking, strategising, and approaching problems the way designers would [25 - 27]. Essentially, it is a holistic approach that aims for innovation [28, 29] using designer's sensibility to satisfy user requirements based on technological feasibility with the possibility of making a practical business strategy, creating a prospective market, and adding value to customers [29]. This research added to the work of Sobanski [1] and Visser [2] to build more on the lean assessment tools focusing more on the internally related structure. Sobanski [1] developed lean assessment tools for warehouse maturity assessment from the practices and principles of lean manufacturing, in which some of the lean tools were not applicable in the warehouse environment. Visser [2] considered eight (8) variables modified from the works of Sobanski [1], Shah, and Ward [30] to develop lean tools with consideration to total supply chain and 3PL. These variables were not totally sufficient to judge performances in the warehouse environment. This article developed fifteen (15) lean tools and ninety-five (95) corresponding lean practices in an effective form of data collection methodology *via* self-reporting of data through design thinking and consulting industries archives as a unique contribution. The research hopes to provide insights and contribute to a better understanding of the intricate processes of developing lean warehousing tools that are often overlooked. The paper was structured as follows: firstly, the introduction and literature review were discussed, followed by the presentation of the concept of design thinking as our theoretical framework. Thirdly, the research method used was described, the outline of the details of the case study, and results were presented based on the analyses of the data gathered from the case study. The research was concluded with a discussion of the research implications and future research.

THEORETICAL FRAMEWORK ON DESIGN THINKING

Design thinking refers to the cognitive processes that are manifested in design action [31, 32]. Similarly, Boland & Callopy [33] observed that thinking is not something that happens solely on instinct but generally occurs based on interactions with people *via* tools. Design thinking approach in literature has been defined both as a fuzzy and analytical process for problem solving in business, learning, health, and organisational contexts, among others. However, there is an agreement with respect to design thinking as a process applicable to the involvement of both analytical and experimental studies [34, 35]. Design thinking comprises of four essential phases during the process:

Idea Generation

This is the process of generating various alternative solutions to problems. Application of design thinking means building solutions by thinking [29]. Also, in design thinking, the creation of prototypes or models is used as the basis of tools for thinking instead of merely as wants to represent an idea. In the context of lean tool development, models are developed from the input and output data from the warehouse facility.

The Developmental Envelopment Analysis (DEA) is a non-parametric linear programming technique used for the evaluation of decision-making units (DMUs) involving multiple inputs and outputs. It identifies the most efficient DMU and measures the efficiencies of other DMUs based on the deviation from the efficient DMU. Taking into account the challenge of complexities in the warehouse environment, performance is adjudged based on efficiency [24]. DEA is an efficient base model proposed by authors to benchmark warehouse performance metrics as inputs to produce outputs [2, 18, 24]. In this study, similar to Visser [2], different warehouses are considered as DMU, where warehouse performance metrics are used as inputs and KPIs as outputs. This makes the DEA model aninput-based model [2]. DEA model is also known as the CCR model to obtain values for weighted inputs Vi (i = 1, 2, 3…..m) and weighted outputs Ui (i = 1, 2, 3…s) with the objective of finding the solutions to weights Vi and Ui that maximize the ratio of DMU while satisfying the input and output ratio constraints of less than 1.

Let the DMU$_j$ be evaluated on any trial designated as DMU$_o$ (where o = 1, 2…n), and then this model is presented as Equations (1-4).

$$(FPo) Max\theta = \frac{u1y1o + u2y2o + \cdots + usyso}{v1x1o + v2x2o + \cdots + vmxmo} = \frac{\sum_r u_r y_{ro}}{\sum_i v_i x_{io}} \qquad (1)$$

$$\text{Subject to } \frac{u_i y_{ij} + \cdots + u_s u_{sj}}{v_i x_{ij} + \cdots + v_m x_{mj}} \le \frac{\sum_r u_r y_{rj}}{\sum_i v_i x_{ij}} \le 1 \; for \; j = 1 \ldots n \qquad (2)$$

$$v_1, v_2 \ldots v_m \ge 0 \; for \; i = 1 \ldots .m \qquad (3)$$

$$u_1, u_2 \ldots u_s \ge 0 \; for \; i = 1 \ldots .s \qquad (4)$$

Where 0 is the objective function value that maximizes the ratio of DMU$_o$, which is also called the relative efficiency score, v_i is the weight for input I, u_r is the weight for output r, x_{io} is the value for input x of DMU$_o$, and y_{ro} is the value for output y of DMU$_o$.

Balancing Analysis and Intuition

Martin [26] suggested that design thinking could be achieved by balancing analytical and intuitive thinking. On one hand, analytical thinking involves past-and-proven or tried-and-tested data, rigour, and quantitative analysis that potentially gain such advantages as repeatability and scalability. On the other hand, intuitive thinking involves creativity and innovation, which Martin [26] referred to as the art of knowing without reasoning. Essentially, according to Lawson and Dorst [36], designers need to make a conscious effort to reach their targeted goal. Moreso, designers, have to also pay attention to their intuition. Intuition has been described as notions and contemplations designers unintentionally compose on a subconscious, unconscious, or preconscious level [37]. In addition, Cross [38] stated that intuitive thinking might be something that designers possess naturally or even derive from their experience, prior learning, and familiar situations. In the context of lean tools development, designers might possibly make design decisions intuitively based on familiar situations, such as previous projects, but at the same time, they would analyse the situation based on what they have done in the past.

Human-Centred: Empathy and Collaboration

Design thinking is human-centered and is achievable through empathy and collaboration [29]. Essentially, Brown [29] argues that one should consider what users need or might need to provide good ergonomics by understanding the cultural context and environment. Furthermore, he argued that the role of users in design has shifted from consumption to participation. Hence, design is too significant to be left in the hands of the designer alone. Rather, it should involve users as part of the collaboration team. In the context of lean tool development, arguably, this needs interaction and collaboration between the lean design team and warehouse operators. This includes the need to understand the concerns and requirements of the warehouse operators. As a result, design thinking enables the development of lean tools that could potentially be useful for warehouse employees.

Creating Innovations

According to Boland and Collopy [33], design novelty is significant. Also, respect for any special conditions or constraints is germane but not limited by them. In applying design thinking, contrasting ideas and constraints need to be exploited to create new solutions [29], thus creating innovations and avoiding being restricted to a specific idea while new ideas are flowing [28, 33]. Another way to assess the opposing constraints on design is to balance desirability, viability, and feasibility [29]. Especially in the context of lean tool development, the intricacy behind the design process cannot be overlooked. Based on its design elements, discourse is unavoidable and the involvement of different groups such as lean experts, organisation management, and warehouse employees. More so, the need for customers and suppliers' aspects integrated into the design process could not be ignored. Also, the impact of the design process on the current technologies and processes in the organisation cannot be underestimated. Every design element is taken into consideration by the lean expert so that the end result, the lean assessment tool itself, could eventually deliver innovation for both the warehouse employee and the organisation. The design process starts with a discourse involving all stakeholders led by a lean expert. This is when warehouse employee states their expectations and needs, while other stakeholders express their concerns related to the design initiative. All these are identified as criteria for the solution. Then lean experts extract information from various data sources (internal and external) in the organisation before outlining the problem objectives. The possibilities of gaps between the problem objectives and the solution criteria are certain. The problem objectives in this context signify the lean expert's entire innovative ideas of how the tool will be developed. The solution criteria suggest some potential constraints that were required and acceptable to the warehouse employees and the management of the organisation. As lean experts seek to reconcile the gap by exploring the problem area and framing the problem, which informs the lean experts when choosing the design principles. At this stage, the solution concepts have to be in line with the problem frame. Additionally, it is possible for the lean experts to work back and forth between the problem frame and solution concepts to finally create tools that satisfy the solution criteria within acceptable time constraints. In subsequent sections, the research method used in this study, an outline of the details of our case study, and a presentation of results based on analyses of the data gathered from the case study shall be discussed.

RESEARCH METHODS – CASE STUDY RESEARCH

The methodology adopted in the study is a case study. This method illustrates how lean warehousing (LW) is used to evaluate the existing warehouse processes in a 3PL company in Nigeria based on productivity, quality, and employee satisfaction. The study also presented lessons learnt and managerial implications of lean warehousing implementation. The case study method was chosen because it offers flexibility in design and implication by allowing both quantitative and qualitative analysis, which are more sensitive to organisation complexities phenomena [39 - 42]. A case study method offers a means of investigating complex and critical functions of the value chain [43, 44]. Another advantage of the method is that it helps make direct observations, collect data in a natural setting, and compare so as to rely on the derived data [43, 45]. In this study, a real-time problem of warehouse inefficiency was considered. The gathered data were based on warehouse operational activities related to suppliers, customers, and employees. The assessment of the warehouse lean maturity focuses on the major functional activities/processes applicable within the warehouse, *e.g.*, inbound, outbound, inventory management, material return, general facility management, and warehouse office administration, in line with lean tools like supplier involvement, customer involvement, statistical process control (SPC), 6S, employee involvement and warehouse specific tools. These tools are valuable for diagnosing and resolving a set of warehousing problems [45, 46].

Design Approach

The design approach in this study is qualitative and quantitative, based on the foundation laid by Shah and Ward [30], Sobanski [1], and Visser [2], where data were gathered through semi-structured interviews with some stakeholders of the organisation. The design team includes lean experts, warehouse employees, and staff at the management level. The interview questions covered areas that were relevant to the process of development of lean warehouse assessment tools. Analysis of the data was done using the qualitative method, a method that is commonly used to analyse an extensive range of data, including interview transcripts and notes on observations [47 - 49]. The lean tools developed were further verified and validated by working within multiple warehousing facilities, each in various stages of lean implementation with unique characteristics to generalise the lean assessment tools developed. The lean tools were refined and operationally defined through onsite analysis and multiple assessors' perspectives. The developed assessment tools utilise a combination of nominal, ordinal, and interval evaluation items, scaled to measure the varying levels of implementation of each of the lean tools and practices in the warehouses. The operationally defined and score evaluation items were aggregated to determine facility lean maturity level so as to identify areas of improvement and provide usable feedback.

Instrumentation

The lean assessment tools and corresponding lean practices comprehensively measure the lean maturity of the warehouse regarding implementation. There are fifteen (15) lean tools and ninety-five (95) corresponding lean practices developed into the framework in the following major operational areas:

i. Inbound operation: These are operations related to material receiving, sorting, checking, stocking, and putting away for inventory purposes.
ii. Outbound operation: These are picking, packing, loading, and shipping processes of materials from inventory to the customers.
iii. Inventory control operation: These are inventory accuracy related to quantity verification, maintenance of stock locations, slotting, and overall facility inventory integrity.
iv. Material return: These are the processes involved with accepting, rejecting, and restocking materials returned from customers/trade.
v. Value-added service operation: These are various tasks performed within warehousing operations such as kitting, packaging, light assembly, and other various tasks to ensure that customers receive products according to specification.

vi. Office administration: This involves managing employees, invoicing, records keeping, human resources, correspondence, meetings, *etc.*

Scale

The scale for the assessment of the degree of implementation of the tools was based on a five-point scale proposed by Shah & Ward [30].

  i. No Implementation
 ii. Little Implementation
iii. Average Implementation
 iv. Extensive Implementation
  v. Full Implementation

Lean Tools (Operational Constructs)

These are a set of warehouse maturity tools for performance assessment. The tools were developed with consideration given to suppliers, customers, and warehouse criteria. Table **1 - 15** present the lean constructs and their corresponding lean practices used for the assessment of the lean maturity in 30 warehouses of the company.

Table 1. Suppliers' involvement construct and measurement items.

Construct	Measurement Item
Sup. 01	Is the warehouse in close contact with suppliers?
Sup. 02	Is there feedback to the suppliers on the inbound delivery performance?
Sup. 03	Is there any attempt to establish long-term relationships with suppliers?
Sup. 04	Is there any standardized supplier certification method?
Sup. 05	Is there a provision of information on inventory position to suppliers?
Sup. 06	Is there provision for efficient material handling and flow by using appropriate means?

Table 2. Customers' involvement construct and measurement items.

Construct	Measurement Item
Cust. 01	Is the warehouse in close contact with customers?
Cust. 02	Is there feedback from the customers on the outbound delivery performance?
Cust. 03	Is there any active involvement in current and future value added logistics offerings by the customers?
Cust. 04	Is there frequent sharing of current and future demand information with the planning department?

Table 3. Statistical process control and measurement items.

Construct	Measurement Item
SPC. 01	Are processes on the shop floor being managed with SPC?
SPC. 02	Is there any use of statistical techniques to reduce process variance?
SPC. 03	Is there any chat and tool to measure productivity on the shop floor?
SPC. 04	Are there any standardized problem solving techniques to identify the root causes of problems?
SPC. 05	Are there any process capability studies before introducing new logistics processes?
SPC. 06	Is the current and desired state of all value streams mapped?

Table 4. 6S Involvement construct and measurement items.

Construct	Measurement Item
Select 01	Are unnecessary, unused and defective items, accessories, materials, equipment, tools, storage bins, and others, removed?
Select 02	Are all unneeded and old information in the working area removed? (Bulletins, work instructions, orders, target deployments, working results, *etc.*)
Sort 01	Are things at the workstation arranged by usage frequency and according to ergonomic aspects?
Sort 02	Are there markings within the working areas?
Shine.01	Are workstations and machines properly cleaned and therefore free of dirt and tidy?
Shine 02	Is the working environment free of dirt?
Shine 03	Are there cleaning schedules and/or checklists (are in the cleaning schedules/checklists all relevant tasks defined by name, description, and scheduled?)
Shine 04	Are all relevant cleaning equipment available and at the designated places (cleaning board, cleaning trolley, cleaning cupboard, *etc.*)?
Stand 01	Are there only the corresponding things at the marked places? (floor spaces, shelves, working spaces, cupboards)
Stand 02	Are the associates of all shifts informed about the targets and procedures of 5S? (5S standard process, checklist cleaning rules, process confirmation, *etc.*)
Stand. 03	Are the existing cleaning schedules visualised and verifiably used daily? (scheduling and execution)
Sust. 01	Are defined standards met and continuously improved? (Process confirmations, filled-in task cards, *etc.*)
Sust. 02	Are the established boards up to date, does the warehouse work consequently and sustainably with relevant tools (OPL, milestone plans, VSM/VSD, problem solving method, *etc.*)?
Sust. 03	Is a regular monthly 5S self-assessment executed by operational/warehouse manager?
Safety 01	Are safety devices (handrails, balustrades, security devices, *etc.*) available and ok?
Safety 02	Is the allowed maximum stacking height/weight of transport and loading devices met?
Safety 03	Are fire extinguishers available/at designated places? (only relevant when fire extinguisher movable)

Table 5. Employee involvement construct and measurement items.

Construct	Measurement Item
Empl. 01	Are shop-floor employees key to a problem solving team?
Empl. 02	Do the shop -floor employees give suggestions to problems?
Empl. 03	Do shop floor employees lead process improvements efforts?
Empl. 04	Do shop-floor employees saddle with decision-making?

Table 6. Warehouse process confirmation construct and measurement items.

Construct	Measurement Item
Inbound 01	Are new stock received recognized into ERP (Enterprise resource planning) within an acceptable period?
Inbound 02	Are there discrepancies that were not reported or solved?
Inbound 03	Are the damaged products in receiving area clearly identified?
St.& Hand 01	Are all locations properly labelled?
St.& Hand 02	Is the storage condition measured, documented, and deviation addressed?
St.& Hand 03	Is repacking according to work instructions done properly?
Outbound 01	Are items being picked and packed ready for transport as per KPI?
Outbound 02	Are deliveries properly packed and labelled?
Outbound 03	Are there any pending return items?
Qual. 01	Are all non-conforming items properly identified and labelled?
Qual. 02	Are stock discrepancies analysed and addressed?
Qual. 03	Is the scrapping process done properly and documented?
Qual. 04	Annual or permanent stocktaking completed?
Qual. 05	Contract for services self-audit completed?

Table 7. Warehouse structured communication construct and measurement items

Construct	Measurement Item
SC 01	Is there a clearly defined communication route with regards to agenda, participants, contents, duration, and frequency per meeting?
SC 02	Is there a display of standardized communication schedule on the CIP board, pentagon, or other official information board?
SC 03	Is there usage of CPI to regulate actions to be done?
SC 04	Is there involvement of all associates by ensuring information flow top-down and bottom-up through meeting schedule?
SC 05	Is there design and execution of an internal customer survey?

(Table 7) cont.....

Construct	Measurement Item
SC 06	Is feedback shared with associates for correction action in PCDA format and closure of PDCA loop?

Table 8. Warehouse risk identification implementation construct and measurement items.

Construct	Measurement Item
RI 01	Is risk operation identified according to impact?
RI 02	Is there a structured method or visuals for the identification of risk (Ishikawa)?
RI 03	Is identified risk included in the work instructions, finding identified defined in CPI format, and addressed to the right person?
RI 04	Is the identification of corrective measures written and followed on PCDA format with responsibilities and deadlines defined?
RI 05	Is there a successful implementation of measures identified?
RI 06	Is feedback shared with associates for correction action in PCDA format and closure of PDCA loop?

Table 9. Warehouse escalation management construct and measurement items.

Construct	Measurement Item
EM 01	Is there a proper definition of the emergency measures and putting in place a reaction system for each process?
EM 02	Is escalation reported to the responsible person (*e.g.,* blue-collar shift-leader, warehouse manager, management)?
EM 03	Any assurance that all KPIs and the intra-days performance escalation triggers are properly defined as well as escalation tasks, responsibilities, hierarchies, and timelines?
EM 04	Is the identification of corrective measures written and followed on PCDA format with responsibilities and deadlines defined?
EM 05	Is there any identification of findings in OPL format and addressed to the correct person?
EM 06	Any structure in place for a quick reaction system for picking and packing to monitor performance in short time slots based on order volume and planned performance?
EM 07	Is there any required capacity done based on the forecast and real productivity per man hour?
EM 08	Ensuring a capacity plan is transparent, and capacity figures are communicated to all responsibilities

Table 10. Warehouse process management implementation construct and measurement items.

Construct	Measurement Item
PM 01	Is there any establishment and usage of mandatory KPIs to be measured at least monthly as reported to the relevant department?
PM 02	Are all monthly KPIs (inbound and outbound quota, productivity, customer claim, and warehouse cost) results displayed on the performance board for visualization?

(Table 10) cont.....

Construct	Measurement Item
PM 03	Any assigned findings from review addressed in OPL format and establishing of regular meeting time, participant and agenda?
PM 04	Is there documentation of daily delivery performance as well as tracking and prioritization of deviation to be displayed on the performance board?
PM 05	Are corrective measures defined in PDCA format, deadline, and responsibility assigned as well as documentation of follow up and corrective measures?
PM 06	Any design and use of active cockpit to monitor KPIs?

Table 11. Warehouse PDCA implementation construct and measurement items.

Construct	Measurement Item
PDCA 01	Is there any definition of progress indicators showing various stages as well as agreed times for updating progress topics?
PDCA 02	Any documentation of implemented action points, responsibilities, and deadlines from PCDA?
PDCA 03	Any regular periodic review of PDCAs and documentation of outcomes?

Table 12. Warehouse waste analysis implementation construct and measurement items.

Construct	Measurement Item
WA 01	Any established and well defined waste analysis method and template for use in a warehouse?
WA 02	Any documentation of the use of waste analysis template and definition of various responsibilities?
WA 03	Any identification of findings in OPL format and addressed to the right personnel?
WA 04	Is there conduction of periodic process-focused waste analysis (good receiving, put-away, picking, packing, and dispatch) as well as periodic training of shop floor employees?
WA 05	Is there a PDCA framework with roles and deadlines for defining, implementing, and documenting corrective measures, progress action, and follow through?

Table 13. Warehouse problem solving implementation construct and measurement items.

Construct	Measurement Item
PS 01	Definition of systematic problem solving approach and root cause analysis as well as provision of problem solving chats?
PS 02	Identification and implementation of corrective measures, written and followed in PDCA format with responsibilities and deadlines assigned
PS 03	Usage of problem solving standard approach in all value stream areas?

Table 14. Warehouse work instruction implementation construct and measurement items.

Construct	Measurement Item
WI 01	Making written work instruction available on the information board in an easy-to-grasp format for all relevant processes?

(Table 14) cont.....

Construct	Measurement Item
WI 02	Design and use of visuals and labels for clarity?
WI 03	Documentation of training plan and schedule for new associates and skills update for old associates?
WI 04	Review and sign off respective schedule, training plan, and actual training?

Table 15. Warehouse visualization implementation construct and measurement items.

Construct	Measurement Item
VI 01	Availability of up dated warehouse layout including functional areas?
VI 02	Marking of areas to indicate the position of materials and machines?
VI 03	Indicating pathways for pedestrians or machines to avoid collision?
VI 04	Marking all inventory outside of racks and shelf locations with a unique identifier including destination and material status?

Verification

The lean assessment tool verification phase was conducted after the design stage. This was carried out to determine the applicability and the correctness of the measures by relating the tools to all the warehouse functions in line with lean design team expectations. The assessments provide objective results in line with expectations. Three warehouse facilities in the Lagos region were assessed two times during 2019 to provide additional insights into the results over time.

Validation

To validate the lean assessment tools, ninety five lean practices were evaluated at 30 warehouse facilities, ensuring measurement outcomes meet expectations at multiple warehouses across industries and geographical regions and ensuring equity among comparisons while identifying future improvements. The corresponding outcome data analyses were conducted using statistical techniques. According to Cronbach [50], validation is a process of collecting evidence to support any conclusion drawn from the test score. Babbie [51] outlines four types of validity: face, criterion, construct, and content validity. The development of the lean assessment tools consists of three phases, namely: theoretical, shop-floor, and piloting phase. The theoretical development phase was completed at the academic level, gathering information from existing literature, tools, and experience. This stage of development addressed the face validity of specific measures, ensuring the reasonableness of potential measures identified to capture a concept. The shop-floor development phase entails gathering input from workers, supervisors, managers, and lean experts for providing insight at a shop-floor level for measurement and feedback. The content validity was addressed by involving multiple levels of inputs from the shop-floor to lean practitioners and academics to ensure the lean constructs and practices were in agreement with the application focus. Furthermore, addressing validity at each stage of the development ensures accurate output from the assessment for statistical analyses. The assessment tool was further validated through a pilot test during onsite shop-floor development. The pilot development addressed criterion related validity by ensuring the lean practices measure the maturity accurately across the whole facilities.

CASE-STUDY BACKGROUND

This case study was undertaken in a Third Party Logistics Company managed as a Private Limited Liability Company. The organisation was established in 1980 as a manufacturing logistics and warehousing company located in the South-West province of Nigeria. The company is made up of thirty (30) functional warehouses spread across the 6 geographical zones; South-West, North-West, North-East, North Central, South-South, and South-East, with a total of seven hundred and fifty (750) staff. The warehouse and depot operations department takes about 70% of the entire labour force of the organisation. The company receives products through the warehouses (inbound) and make deliveries of orders (outbound) on a daily basis. All the warehouses have varying amounts of store keeping units (SKUs). Fig. (**1**) shows the value chain of the warehousing processes. Value chain gives the idea that the process is composed of other sub processes, each with input, transformation activities, and their respective output [52, 53]. It is a systematic approach to examine the development of competitive advantage [57]. The stages that represent the value chain of warehousing are described as follows:

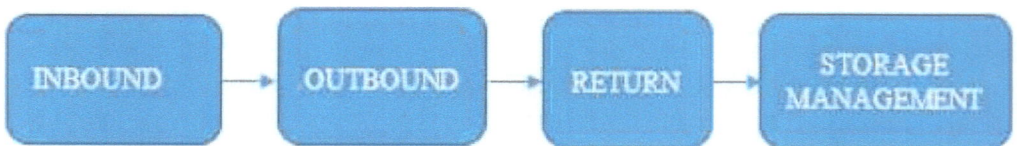

Fig. (1). Value chain of the warehousing.

  i. Inbound
 ii. Outbound
iii. Return
iv. Storage Management

The warehouses are homogeneous with similar inputs and outputs. The warehouses differ in terms of numbers and types of products stored, number of employees both full time employees (FTE) and contracts, size of the warehouse, number of daily customer orders, and layout.

Business Case

In order to maintain good customer-to-customer relationships through organizational management, the efficiency of the warehousing process needs to be improved by assessing the lean maturation process. The management decided to engage lean six sigma practitioners to evaluate the current warehousing processes of the company. The Lean Six Sigma team is composed of a researcher specialized in quality management and a senior lecturer at the University who is also Black Belt certified. The team engaged in the company's document review, one-on-one interview, and questionnaires attestation with some selected staff from the warehousing department. These were further supported by self-observations on the shop floor for a period of six months to understudy the warehousing processes.

Data Collection

Data used were gathered from both primary and secondary sources. The primary data collected were *via* structured interviews, questionnaires, and observations from a Logistics Company operating in Nigeria as a 3PLService provider. Two lean specialists developed a list of measurement items, which were aggregates of assessment tools (Tables **1-15**) used in the warehouse environment. The assessment tools were limited to standard basic lean maturity tools. Questionnaires on lean maturity implementation were sent to thirty (30) warehouses of the company to be attested by the Warehouse Managers and two (2) other Assistant Warehouse Managers. The researcher carried out self-observations in the form of internships for a period of six (6) months to understudy warehousing operations. The secondary data were collected from the literature, internet, and from Bosch archive for lean maturity assessment tools

The following four (4) inputs and three (3) outputs were used:

Labour hours: The total annual man hour for all direct full time employees (FTEs) and contract staff directly involved in all of the inbound and outbound warehouse activities. Inbound activities include offloading, receiving products into the warehouse storage, update of inventory data on ERP, and outbound activities include all activities involved in picking, packing, and shipping of products to specific customers.

Warehouse Space: Total warehouse space used inbound activities, outbound activities, stock, and storage. Floor space is a function of the area required for daily activities in the warehouse, including a number of SKUs in the storage locations. Warehouse space is referred to as the total floor area per square feet. Consideration is given to the length, breath, and height as part of the warehouse space. The average space allocation by functional area as a percentage of the total warehouse space in all the warehouses is presented in Table **16**.

Table 16. Percentage warehouse space.

Area	Percentage space
Storage Space	51
Encumbrances (staging, aisles)	37
Office space	5
Facilities (rest rooms, canteen)	3
Value added services	4
Total warehouse space	100

Level of Automation: This is the total annual expenses of automation accrued in support of each warehouse operation. These costs include all startup technology and annual recurring costs. Details are presented in the following categories:

i. Hardware: Radio frequency (RF) equipment, computers, CCTV, cameras, printers, scanner, internet servers
ii. Software license fees
iii. Infrastructure installation
iv. Warehouse management integration system and support applications
v. Testing and training
vi. Project management development and implementation

Number of Stock Keeping Units (SKUs): This is the total number of assortments/products stored per annum in the storage locations in the warehouse. The size of the SKUs in a warehouse impacts the potential output rate. The size of assortment is measured through the amount of SKUs using an 8-point scale of ≤500; 500-1000; 1000-5000; 5000-10,000; 10000-20000; 20,000-50,000; 50,000-100,000 and ≥100,000. The summary of the warehouse input is presented in Table **17**.

Table 17. Summary of warehouse inputs.

Warehouse	Labour	Space	Automation	SKUs
W1	134992	500400	903000	1000-5000
W2	68640	202500	413000	1000-5000
W3	100672	312300	650000	1000-5000
W4	224576	97200	358400	1000-5000

(Table 17) cont.....

Warehouse	Labour	Space	Automation	SKUs
W5	183744	180000	324100	1000-5000
W6	377520	360000	646800	1000-5000
W7	82368	172959	132000	1000-5000
W8	75504	76500	318000	1000-5000
W9	41184	54000	227500	1000-5000
W10	43472	180,000	200900	1000-5000
W11	41184	99000	126000	1000-5000
W12	45760	63000	151200	1000-5000
W13	167024	322336	268300	1000-5000
W14	160160	92110	286300	1000-5000
W15	114400	187331	175000	1000-5000
W16	77792	72000	87500	1000-5000
W17	304304	196200	641200	1000-5000
W18	269984	360000	305200	1000-5000
W19	247104	184000	384600	1000-5000
W20	46750	31910	353500	1000-5000
W21	79904	278100	178119	1000-5000
W22	34650	576000	100279	1000-5000
W23	114041	365110	472118	1000-5000
W24	74316	30600	457800	1000-5000
W25	157520	31500	140000	1000-5000
W26	25887	99000	320500	1000-5000
W27	25814	411010	243523	1000-5000
W28	40205	108000	245000	1000-5000
W29	62458	64800	420000	1000-5000
W30	841753	33026	372400	1000-5000

The following three outputs were used:

Shipping Volume: This determines the quantity of orders fulfilled per annum. Warehouse volume is usually measured in units, boxes, pallets, *etc.*

Percentage of Error Free Orders Shipped: This is the ratio of good orders filled completely and in time to the total orders shipped. It is the annual percentage of error free orders shipped.

Space Utilization: This is calculated as the total product cubic displacement divided by the total warehouse cubic space. Space utilization is considered in three dimensions to examine the impact of warehouse height on its efficiency. It measures the total cubic meter utilized. Table **18** presents the summary of the warehouse outputs.

Table 18. Summary of the warehouse outputs.

Warehouse	Order Shipped	% EFO	Cubic feet Utilized
W1	1992656	1073390	8095360
W2	1844270	1659843	3307500
W3	1449542	1304587	5130048
W4	980411	882370	1747384
W5	911760	820584	3570000
W6	1846300	1661670	6451200
W7	332800	2995200	3164002
W8	977600	879840	1416100
W9	561600	505440	919800
W10	1050616	945554	3920000
W11	345600	311040	2217600
W12	589680	530712	1348480
W13	487640	438876	6325323
W14	546916	492224	1664933
W15	433328	389995	3321994
W16	484667	436201	1155840
W17	750126	675108	3662400
W18	606528	545875	6977600
W19	707098	636389	3247531
W20	448098	403288	162100
W21	1076543	968889	1920100
W22	404505	364054	2193800
W23	374257	336831	3953048
W24	99566	89609	3150000
W25	189600	170640	2019346
W26	693760	624384	2419620
W27	1131662	1018496	1633100
W28	366240	329616	854000

(Table 18) cont.....

Warehouse	Order Shipped	% EFO	Cubic feet Utilized
W29	448000	403200	3166178
W30	614600	553140	1785076

RESULTS

This section presents the results of the lean maturity assessment of the 30 warehouses by the implementation of Lean warehousing assessment tools.

Lean Maturity Assessment

The assessment of all the warehouses was carried out by the warehouse managers with the support of the warehouse supervisors based on the developed lean tools. The responses to the tools were based on the five point scales presented in Table **19**. The degree of implementation of lean differs from one warehouse to the other. The maturity evaluation is to determine the performance of the thirty warehouses in terms of warehouse productivity, quality, and employee satisfaction. The design of the operational constructs (assessment tools) was done to cover the six functional areas of warehousing operations. The warehouse operations were assessed against each of the lean practices of the lean tools. The ranking of the warehouse performance was determined using the highest lean maturity scores of A-C. The outcomes of the assessment and statistical analyses were presented in Table **20** and **21**, respectively. Fig. (**2**) represents the statistical performance ranking of the warehouses.

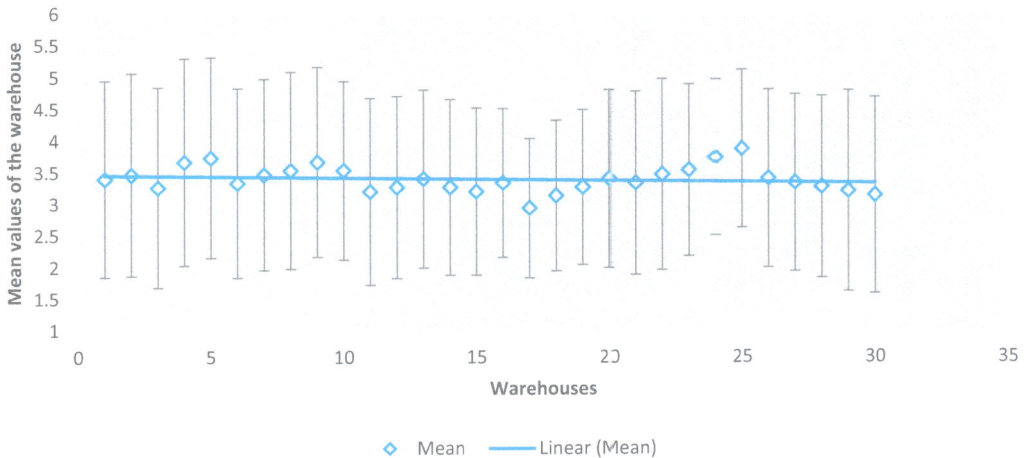

Fig. (2). Error bar from the statistical analysis of the warehouses

Table 19. Lean maturity assessment scale.

Scale	Description	Maturity Score	Rank
0	No Implementation	$\leq 40\%$	E
1	Little Implementation	$\geq 40 - 60\%$	D
2	Average Implementation	$\geq 60 - 80\%$	C
3	Good Implementation	$\geq 80 - 90\%$	B
4	Full Implementation	$\geq 90 - 100\%$	A

The developed lean tools are defined as follow:

i. Supplier Related (SUP): The lean practice is aimed at a good relationship between suppliers and warehouse operators in terms of supplier's feedback, Just-in-Time delivery, and supplier's development.

ii. Customer Related (Cust): The lean practice is aimed at a good customer-client relationship in terms of customer feedback and Just-in-Time delivery.

iii. Statistical Process Control (SPC): The lean practice is aimed at process monitoring and error tracking. It ensures a flow of error free orders throughout the warehouse.

iv. 6S: This is aimed at the systematic way of ensuring and improving orderliness, cleanliness, visualization, and safety in the warehouse.

v. Employee Related: This lean practice is targeted at the employee of the warehouse, both FTE and contract, aimed at employee development and welfare.

vi. Process Confirmation (PC): A lean tool aimed at checking if defined process standards are followed.

vii. Structured Communication (SC): It is a lean tool that ensures that communications are defined in line with the agenda, participant, in accordance with schedules.

viii. Risk Identification (RI): It is a tool to proactively identify the risk of processes, evaluate, and mitigate the risk of the process.

ix. Escalation Management (EM): A lean tool ensures defined intervention limits and communication flow in case of disruptions in daily business. Also, standardized tools to ensure defined actions in case of deviations from the standard.

x. Performance Management (PM): A lean tool monitors both performance and improvement measures to manage deviations from standards on a daily basis.

xi. Plan-Do-Check-Act (PDCA): A lean tool aimed at deviation management. It is a cycle of process steps that enables continuous improvements when implemented consistently.

xii. Waste Analysis (WA): This is a lean tool aimed at elimination of non-valu-
-adding activities in processes, operations, and resources,

xiii. Problem Solving (PS): A lean tool ensures a systematic approach to find
unknown root causes for problems, to define and implement respective
solutions.

xiv. Work Instruction (WI): It is a lean tool that ensures defined standards for
shop floor operations in a step-by-step written form.

xv. Visualization (VI): It a lean tool that ensures transparency of warehouse
processes

Table 20. Maturity assessment outcomes.

S/N	WH	SUP	CUST	SPC	6S	EMPL	PC	SC	RI	EM	PM	PDCA	WA	PS	WI	VI		A	B	C	D	E	HLM	Rank
1	W1	C	B	E	A	B	A	E	C	E	C	D	C	A	A	A	WI	5	2	4	1	3	11	9
2	W2	C	B	E	A	B	A	E	C	E	C	D	A	A	A	A	W2	6	2	3	1	3	11	9
3	W3	C	B	E	A	B	D	E	C	E	C	D	A	A	A	A	W3	5	2	3	2	3	10	22
4	W4	A	B	E	A	B	A	E	C	E	B	D	A	A	A	A	W4	7	3	1	1	3	11	9
5	W5	A	B	E	A	B	A	E	C	E	B	C	A	A	A	A	W5	7	3	2	0	3	12	3
6	W6	A	B	E	A	C	C	E	C	E	C	C	C	A	A	A	W6	5	1	6	0	3	12	3
7	W7	A	B	E	A	C	B	E	C	E	B	C	C	A	A	A	W7	5	3	4	0	3	12	3
8	W8	A	B	E	A	B	B	E	C	E	B	D	B	A	A	A	W8	5	5	1	1	3	11	9
9	W9	A	B	E	A	B	B	E	C	D	B	D	A	A	A	A	W9	6	4	1	2	2	11	9
10	W10	A	B	E	A	B	B	E	C	D	B	D	B	A	A	B	W10	4	6	1	2	2	11	9
11	W11	C	B	E	A	B	B	E	D	D	C	E	B	A	A	B	W11	3	5	2	2	3	10	22
12	W12	C	B	E	A	B	B	E	D	C	C	E	B	B	A	A	W12	3	5	3	1	3	11	9
13	W13	B	C	E	A	B	A	D	D	C	C	E	B	B	A	A	W13	4	4	3	2	2	11	9
14	W14	B	C	D	B	B	A	E	C	D	E	B	B	A	A	A	W14	3	5	2	3	2	10	22
15	W15	B	C	D	B	B	A	E	D	C	D	E	B	B	A	B	W15	2	6	2	3	2	10	22
16	W16	B	C	D	B	B	A	D	C	C	D	E	B	B	A	B	W16	2	6	3	3	1	11	9
17	W17	B	D	D	C	C	C	D	C	E	D	D	B	B	A	B	W17	1	4	4	5	1	9	28
18	W18	B	B	D	C	C	C	D	C	E	D	D	A	B	A	B	W18	2	4	4	4	1	10	21
19	W19	B	B	E	C	C	C	D	C	E	B	C	A	B	A	B	W19	2	5	5	1	2	12	3
20	W20	B	B	E	A	C	C	D	D	E	B	C	A	A	A	B	W20	4	4	3	2	2	11	9
21	W21	B	B	E	B	C	B	E	D	E	B	C	B	A	A	A	W21	3	6	2	1	3	11	9
22	W22	B	B	E	A	B	B	E	C	B	C	B	A	A	A	A	W22	4	6	1	1	3	11	9
23	W23	B	C	E	B	B	A	E	C	D	B	C	B	A	A	A	W23	4	5	3	1	2	12	3
24	W24	A	C	C	B	B	A	E	C	D	B	C	B	A	A	A	W24	5	4	4	1	1	13	1
25	W25	A	B	C	B	B	A	E	C	D	B	C	B	A	A	A	W25	5	5	3	1	1	13	1
26	W26	A	B	C	C	C	A	E	D	E	C	C	C	A	A	A	W26	5	1	6	1	2	12	3
27	W27	C	B	D	B	C	A	E	D	E	B	C	C	A	A	A	W27	4	3	4	2	2	11	9
28	W28	B	B	D	C	A	E	D	D	B	E	C	A	A	A	A	W28	4	3	3	3	2	10	22
29	W29	B	D	E	A	B	B	E	D	D	B	E	C	A	A	A	W29	4	4	1	3	3	9	28
30	W30	C	D	E	A	B	B	E	D	D	B	E	C	A	A	A	W30	4	3	2	3	3	9	28

Table 21. Statistical analysis of the warehouses.

WAREHOUSE	N	RANGE	MIN	MAX	SUM	MEAN	STD ERROR	STD DEV	SKEW	STD ERR	KURTOSIS	STD ERROR
WH 1	15	4,00	1,00	5,00	51,00	3,40	40000	1,54919	-516	580	-1,163	1,121
WH 2	15	4,00	1,00	5,00	52,00	3,4667	41250	1,59762	-537	580	-1,253	1,121
WH 3	15	4,00	1,00	5,00	49,00	3,2667	40786	1,57963	-256	580	-1,474	1,121
WH 4	15	4,00	1,00	5,00	55,00	3,6667	42164	1,63299	-354	580	-960	1,121
WH 5	15	4,00	1,00	5,00	56,00	3,7333	40786	1,57963	-999	580	-485	1,121
WH 6	15	4,00	1,00	5,00	50,00	3,3333	38627	1,49603	-366	580	-966	1,121
WH 7	15	4,00	1,00	5,00	52,00	3,4667	3631	1,50555	-551	580	-793	1,121
WH 8	15	4,00	1,00	5,00	53,00	3,5333	380048	1,55226	-799	580	-882	1,121

(Table 21) cont.....

WAREHOUSE	N	RANGE	MIN	MAX	SUM	MEAN	STD ERROR	STD DEV	SKEW	STD ERR	KURTOSIS	STD ERROR
WH 9	15	4,00	1,00	5,00	55,00	3,6667	3711	1,49603	-815	580	-791	1,121
WH 10	15	4,00	1,00	5,00	48,00	3,5333	36253	1,40746	-788	580	-624	1,121
WH 11	15	4,00	1,00	5,00	49,00	3,2	35813	1,47358	-396	580	-1,253	1,121
WH 12	15	4,00	1,00	5,00	48,00	3,2667	34087	1,43759	-539	580	-931	1,121
WH 13	15	4,00	1,00	5,00	49,00	3,4	30342	1,40408	-479	580	-918	1,121
WH 14	15	4,00	1,00	5,00	50,00	3,2667	28396	1,38701	-365	580	-1,125	1,121
WH 15	15	4,00	1,00	5,00	44,00	3,2	30654	1,32017	-421	580	-1,001	1,121
WH 16	15	4,00	1,00	5,00	47,00	3,3333	31573	1,17514	-451	580	-555	1,121
WH 17	15	4,00	1,00	5,00	49,00	2,9333	36253	1,09978	148	580	-676	1,121
WH 18	15	4,00	1,00	5,00	51,00	3,1333	37374	1,18723	4	580	-791	1,121
WH 19	15	4,00	1,00	5,00	50,00	3,2667	38873	1,2228	-587	580	-88	1,121
WH 20	15	4,00	1,00	5,00	52,00	3,4	35006	1,40408	-479	580	-918	1,121
WH 21	15	4,00	1,00	5,00	53,00	3,3333	31573	1,44749	-676	580	-854	1,121
WH 22	15	4,00	1,00	5,00	56,00	3,4667	32170	1,50555	-796	580	-810	1,121
WH 23	15	4,00	1,00	5,00	51,00	3,5333	36253	1,35576	-776	580	-293	1,121
WH 24	15	4,00	1,00	5,00	58,00	3,7333	36078	1,2228	-765	580	108	1,121
WH 25	15	4,00	1,00	5,00	51,00	3,8667	37118	1,24595	-992	580	392	1,121
WH 26	15	4,00	1,00	5,00	50,00	3,4	40473	1,40408	-300	580	-850	1,121
WH 27	15	4,00	1,00	5,00	49,00	3,3333	40079	1,39728	-329	580	-970	1,121
WH 28	15	4,00	1,00	5,00	51,00	3,2667	47646	1,43759	-206	580	-1,261	1,121
WH 29	15	4,00	1,00	5,00	48,00	3,20	40473	1,56753	-252	580	-1,583	1,121
WH 30	15	4,00	1,00	5,00	47,00	3,1333	40079	1,55226	-122	580	-1,548	1,121

WH E10	WH E11	WH E12	WH E13	WH E14	WH E15	WH E16	WH E17	WH E18	WH E19	WH E120
15	15	15	15	15	15	15	15	15	15	15
3.5333	3.2000	3.2667	3.4000	3.2667	3.2000	3.3333	2.9333	3.1333	3.2667	3.4000
1.40746	1.47358	1.43759	140408	1.38701	1.32017	1.17514	1.09978	1.18723	1.22280	1.40408
.297	240	228	199	235	261	248	202	167	214	199
.149	132	143	127	153	152	152	202	163	141	127
-297	-240	-228	-199	-235	-261	-248	-167	-167	-214	-199
.297	240	228	199	235	261	248	202	167	214	199
001[c]	020[c]	034[c]	114[c]	026[c]	007[c]	014[c]	101[c]	200[0,0]	064[c]	114[c]
.115	304	359	530	327	217	267	510	735	439	530
000	000	000	000	000	000	000	000	000	000	000

(Table 21) cont.....

WH E21	WH E22	WH E23	WH 124	WH E25	WH E26	WH E27	WH E28	WH E29	WH E30
15	15	15	15	15	15	15	15	15	15
3.3333	3.4667	3.5333	3.7333	3.8667	3.4000	3.3333	3.2667	3.2000	3.1333
1.44749	1.50555	1.35576	122280	1.24595	1.40408	1.39728	1.43759	1.56753	1.55226
.277	305	235	186	218	212	150	162	228	178
.147	154	140	150	182	212	128	144	178	167
-277	-305	-235	-186	-218	-206	-150	-162	-228	-178
.277	240	235	186	218	212	150	162	228	178
003^C	001^C	026^C	170^C	052^C	068^C	200^C	$200^{o,d}$	034^0	$200^{o,d}$
.164	098	328	610	412	448	839	771	359	663
000	000	000	000	000	000	000	000	000	000

Lean Maturity Evaluation Discussion and Lessons Learnt

From Table **20**, W_{24} and W_{25} have the highest lean maturity rank of 13. The level of consistency in the ranking was verified by the statistical analysis carried out (Table **20** and **21**). Compared to other warehouses, $W_1 - W_{15}$, $W_{20} - W_{23}$ as well as $W_{26} - W_{30,}$ $W_{24,}$ and W_{25} have low values of standard deviation (1.2280 and 1.24595, respectively). This implies that the ranking procedure, which produces W24 and W25 as the lean tool with the highest lean maturity rank, was consistent. The lower the standard deviation from the mean value, the higher the level of consistency in the ranking and *vice versa*. However, the lowest value of standard deviation was found in W_{17} (1.09978). This implies that the ranking procedure, which produces W17 as the lean tool with the lowest lean maturity rank, was consistent. The standard deviations in relation to the mean values for the 30 warehouses are depicted using the error bar presented in Fig. (**2**). The longer the error bar, the lower the level of consistency in the lean maturity ranking and *vice versa*. Overall, the magnitude of the mean ranges between a minimum value of 3.1333 and a maximum value of 3.7333. On the other hand, the values of the standard deviation range between a minimum value of 1.09978 and a maximum level of 1.632999 for the 30 warehouses considered. The fact that the values of the mean and standard deviation were found within a similar range of values indicates that the level of consistency in the ranking of the maturity level of the lean tool was high. This lends credence to the fact that the process of ranking the warehouse to select the best in terms of lean maturity was very reliable.

Majority of the lean maturity assessment tools were fully implemented according to the assessment scores. It was observed that the two warehouses rarely developed good practices with their suppliers in terms of Just-in Time delivery of goods and feedback to suppliers. It was also noticed that the two warehouses adhered fully to defined process standards. Complete implementation of systematic approaches to finding the root causes to problems within the warehouse, step-by-step approaches to shop floor operations were well defined in a written format, and transparency in the warehouse processes were fully implemented. It is also observed that W_{24} and W_{25} were doing well in the area of customer relations. There were improvements in orderliness, cleanliness, visualization, and safety in the warehouses. Employee relations were also given extensive attention. Performance management in terms of improvement and performance measures not to deviate from standards were well defined and extensively implemented. The deficiencies of the two warehouses were observed in the area of structured communication and escalation management. In the area of structured communication, the display of standardized communication schedules on the CIP board or official information board is lacking. The use of an open point list (OPL) to register actions to be executed is also lacking with no involvement of associates for information dissemination. The feedback shared with associates for correction action in the Plan-Do-Check-Act (PDCA) format was out of place. Finally, no knowledge of the close of the PDCA loop was demonstrated. In the area of escalation management, there was slight implementation. Identification of corrective measures, written and followed on PDCA format with responsibilities and deadlines were poorly defined. A quick reaction system to monitor performance for picking and packing in the short time slot, based on volume ordered, is lacking. Six warehouses (W_5, W_6, W_7, W_{19}, $W_{23,}$ and W_{26}) came second in the lean maturity ranking with an HLM score of 12. According to the maturity assessment outcomes, the majority of the six warehouses performed excellently well in supplier related, 6S, problem solving, work instruction, and visualization. W_5 implemented complete process confirmation and waste analysis compared to W_6 and W_7. Comparing W_{19} to the others ranked two in maturity, W_{19} implemented lean with a score in the range of $\geq 80 - 90\%$, but not to others. The warehouse was different in the area of provision of information on inventory position to suppliers. In terms of customer related, W19 is deficient in the involvement of current, and future value added logistics offerings. In the assessment of lean tools 6S, process confirmation, employee related, risk identification, and PDCA, W_{19} partially implemented them compared to W_5, W_6, and W_7. The six warehouses implemented lean with scores in the range of $\geq 40 - 60\%$ in escalation management.

Thirteen (13) warehouses (W_1, W_2, W_4, W_8, W_9, W_{10}, W_{12}, W_{13}, W_{16}, W_{20}, W_{21}, $W_{22,}$ and $_{27}$) were ranked third in the lean maturity assessment with an HLM score of 11. Six warehouses (W_3, W_{11}, W_{14}, W_{15}, $W_{18,}$ and W_{28}) were ranked forth with an HLM score of 10. These warehouses ranked third and fourth positions in the lean maturity assessment, respectively

Three warehouses (W_{17}, $W_{29,}$ and W_{30}) have the least lean maturity rankings with an HLM score of 9. From the maturity assessment outcomes, Warehouses 17, 29, 30 show little or no implementation of the lean tools. No implementation of statistical process control (SPC), structured communication, and PDCA. This can be traced to lack of use of statistical techniques to reduce process variations, lack of process capability studies before introducing new logistics processes, as well as lack of defined progress indicators to show any stage of progress and agreed time for updating progress topics. There was no documentation of implemented action points, responsibilities, and deadlines from PDCA. Little implementation was observed in the area of customer related, risk identification, and escalation management. The three warehouses had a full implementation of problem solving, work instructions, and visualization tools. Great attention was given to the systematic problem solving approach and root cause analysis. Corrective measures were identified and implemented according to the PDCA format. Work information was written on the board in an easy to grasp format. Pathways were indicated for pedestrians and machines to avoid collision/accident.

Warehouse Lean Maturity Analysis

This section presents the warehouse performances in terms of warehouse productivity, warehouse quality, and employee satisfaction:

Analysis of Warehouse Productivity

Warehouse productivity focuses on the organisational achievement to meet up with both supplier's and customer's demands with Just-In-Time delivery of customer's goods. In this study, warehouse productivity was judged based on the lean constructs and practices of supplier related, customer related, 6S, process confirmation, structured communication, performance management, plan-d--check-act, problem solving, and work instructions. From the lean maturity assessment outcomes (Table **20**), the warehouse productivity falls within the range of good implementation. It was observed that lean practices improved productivity when structural communication, performance management, and PDCA were implemented. It was discovered that communication in line with the organization's goals may not be effectively defined in accordance with schedules, resulting in poor management performance that required improvement measures.Also, deviations from standards were not properly managed. These findings agreed with the works of Swank [54] and Shah & Ward [30]. The study conducted by De Koster & Warffemius [23] and De Koster & Balk [24] confirmed the possibility of a decrease in warehouse productivity due to the poor structure of the organisation.

Analysis of Warehouse Quality

Warehouse quality focuses mainly on customer's relation and feedback. Also, Just-In-Time delivery of customer's goods and condition of delivery of customer's goods in terms of an error on free order shipped. In this study, warehouse quality was judged based on lean constructs of customer related, statistical process control (SPC), 6S, process confirmation, risk identification, escalation management, performance management, waste analysis, and visualization. From the lean maturity assessment outcomes (Tables 20), the warehouse quality falls within the range $\geq 60 - 80\%$ of average level implementation. The lean maturity in terms of quality declined to $\geq 60 - 80\%$ due to poor implementation of statistical process control, escalation management, and performance management tools. These results agreed with previously carried out works which explained that improved customer focus increases quality maturity, which implies that the percentage of error free orders shipped increases linearly with quality.

Analysis of Warehouse Employee Satisfaction

Warehouse employee satisfaction was based on employee involvement, structural communication, 6S, and visualization tools. The focus of the assessment was to ensure the safety of both workers and equipment, welfare and wellbeing of the employees. This also ensures that employees are part of the stakeholders in terms of decision-making and productivity of the warehouse. From the assessment results, the level of maturity in terms of employee satisfaction falls within the range of average implementation. There was a decrease in employee satisfaction which was traceable to poor implementation of structural communication tools. This had a significant effect on the overall maturity because employee satisfaction is attitudinal [2]. This also agreed with the studies of Hackman & Oldham [57]; Vidal [58]. According to Hackman and Oldham [57], employee satisfaction is influenced by the individual rather than the situation itself. Vidal [58], in his study, made an assertion that not all employees in such organisations can be positive whereas some would be negative while others stay neutral.

Implications of the Study

The use of lean had proved its usefulness in warehouse and logistics environment by this research on lean manufacturing. There was progress in the recent contributions by researchers on the implementation of lean practices in the manufacturing sector. Few of the early researches were directed toward the effect on warehouse performances. Lean practices, when used with lean thinking, provides solutions to real-time problems effectively. This article presented lean maturity in the warehousing environment with direct implications on customers' satisfaction.

This article presents the following theoretical implications:

It encouraged the blending of design thinking with lean. As a result of the recent emergence of design thinking as a full management discipline for research, the blending of lean manufacturing with design thinking offers more opportunities for research in the future. From this research, a synergy was established by the design thinking framework [45].

The managerial implications from the case study include:

The results of the lean assessments made it clear that lean in the warehouse is focused more on the warehouse internally related and customer's related. Customers and employees centred culture is key to lean warehousing [45]. Based on the work of Bhasin & Burcher [59], a lean culture pays attention to making decisions at the lowest level, continuous improvement driven by shop-floor employees, and lean leadership at all levels of the organisation. The implementation of these principles lacks within the warehouses studied thus, compounding to the low lean maturity. To gain more from lean manufacturing, these principles should be adopted by the warehouse as part of its cultural practices [60].

The involvement of industry experts at the master black belt level alongside quality management students was helpful in the development and validation of the lean warehousing tools. Therefore, academic-industry collaboration is recommended. Employee cultural and customer centred focus of lean warehousing, when blended with design thinking, offers a unique combination to solve complex real-time problems.

CONCLUSIONS

The following are the conclusions drawn from the study:

1. The research has successfully developed tools for the assessment of lean maturity in a warehouse environment using the design-thinking method.
2. The effects of the implementation of lean on the warehousing assessment of maturity in terms of productivity, quality, and employee satisfaction are significant.
3. Analyses of the warehouses' performances in terms of productivity show good implementation with a score of ≥80-90%.
4. Analyses of the warehouses' performances in terms of quality and employees satisfaction show an average level of lean implementation with each score of

$\geq$60-80%.

LIMITATIONS AND RECOMMENDATIONS

The following are some of the limitations and recommendations resulting from this study.

1. The research is limited to a case of lean warehousing in a 3PL environment only. It cannot be generalized for a case of the total supply chain.
2. The developed lean constructs are limited to basic lean maturity assessments. The tools can still further be extended to advanced tools.
3. The validation of the developed lean constructs and the corresponding lean practices are limited to empirical validation. They can still be validated by models found in the literature.

CONSENT FOR PUBLICATION

Not applicable.

CONFLICT OF INTEREST

The author declares no conflict of interest, financial or otherwise.

ACKNOWLEDGEMENTS

Declared none.

FUNDING

The authors received no financial support for the research, authorship, and/or publication of this article.

REFERENCES

[1]　E.B. Sobanski, "Assessing Lean Warehousing: Development and validation of a lean assessment tool", In: *Ph.D. thesis* Oklahoma State University: Oklahoma, 2009, pp. 1-290. [Online] Available at, http://citeseerx.ist.psu.edu/viewdoc/download?doi=10.1.1.630.671&rep=rep1&type=pdf

[2]　J.J. Visser, *"Lean in the Warehouse: Measuring Lean Maturity and Performance within Warehouse Environment"*, Master Thesis in Supply chain Management. Rotterdam School of Management, Erasmus University: Rotterdam, pp. 1-802014. [Online] Available at, https://www.erim.eur.nl/fileadmin/user_upload/Lean_in_the_warehouse_-_Jeffrey_de_Visser.pdf [Accessed 2nd February, 2021].

[3]　V. Gupta, R. Jain, M.L. Meena, and G.S. Dangayachi, "Six-Sigma application in tire manufacturing company: A case study", *J Ind Eng Int.,* vol. 14, pp. 511-520, 2018.
[http://dx.doi.org/10.1007/s40092-017-0234-6]

[4]　M.D. Sokovic, K. Pavletic, and K. Pipan, "Quality improvement methodologies: PDCA cycle, RADAR matrix, DMAIC and DFSS", *J Achiev Mater Manuf Eng.,* vol. 43, no. 1, pp. 476-483, 2010.

[5]   C. Roriz, E. Nunes, and S. Sousa, "Application of Lean production principles and tools for quality improvement of production process in a cartoon company", *Procedia Manuf.,* vol. 11, pp. 1069-1076, 2017.
        [http://dx.doi.org/10.1016/j.promfg.2017.07.218]

[6]   P. Dehdari, *"Measuring the Impact of Techniques on Performance Indicators in Logistics Operations",* Ph.D thesis, Karlsruher Instituts für Technologie: Karlsruhe, pp. 1-191 2013. [Online] Available at, file:///C:/Users/daniyania/Downloads/Dehdari_Payam.pdf [Accessed 12th February, 2021].

[7]   J.P. Womack, and D.T. Jones, *Lean Thinking: Banish Waste and Create Wealth in Your Corporation.* Simon and Schuster: New York, 1996.

[8]   M.A. Overboom, D.A.C. Haan, and N.A.A.M. Naus, "Measuring the degree of leanness in logistics service providers: Development of a measurement tool", In: *Proceedings of the 17th International Annual EurOMA Conference: Managing Operations in Service Economies,* R. Sousa, C. Portela, S. Pinto, H. Correia, Eds., European Operations Management Association: Porto., 2010..

[9]   H.B. Gunasekaran, and F.M. Marri, "Improving the effectiveness of warehousing operations: a case study", *Ind. Manage. Data Syst.,* vol. 99, no. 8, pp. 328-339, 1999.
        [http://dx.doi.org/10.1108/02635579910291975]

[10]   A. Hamdan, and K.J. Rogers, "Evaluating the efficiency of 3PL logistics operations", *Int. J. Prod. Econ.,* vol. 113, pp. 235-244, 2008.
        [http://dx.doi.org/10.1016/j.ijpe.2007.05.019]

[11]   A. Johnson, and L. McGinnis, "Performance Measurement in the Warehousing Industry", *IIE Trans.,* vol. 43, no. 3, pp. 220-230, 2010.
        [http://dx.doi.org/10.1080/0740817X.2010.491497]

[12]   I. Abushaikha, L. Salhieh, and N. Towers, "Improving distribution and business performance through lean warehousing", *Int. J. Retail Distrib. Manag.,* vol. 46, no. 8, pp. 730-800, 2018.
        [http://dx.doi.org/10.1108/IJRDM-03-2018-0059]

[13]   C.M. Pereira, R. Anholon, I.S. Rampasso, O.L.G. Quelhas, W.L. Filho, and L.A.S. Eulalia, "Evaluation of Lean practices in warehouses: Analysis of Brazilian reality", *Int. J. Prod. Perform. Manag.,* vol. 70, no. 1, pp. 1-20, 2021.
        [http://dx.doi.org/10.1108/IJPPM-01-2019-0034]

[14]   D.J. Nightingale, and J.H. Mize, "Development of a Lean enterprise transformation maturity model", *International Knowledge Systems Management Journal,* vol. 3, pp. 15-30, 2002.

[15]   A. Mahfouz, *"An Integrated Framework to Assess 'Leanness' Performance in Distribution Centres",* Ph.D. thesis, Dublin Institute of Technology: Dublin, pp. 1-1942011. [Online] Available at, https://arrow.tudublin.ie/cgi/viewcontent.cgi?article=1013&context=busdoc [Accessed on 15th January, 2021].

[16]   T.L. Doolen, and M.E. Hacker, "A review of lean assessment in organizations: an exploratory study of lean practices by electronics manufacturers", *J. Manuf. Syst.,* vol. 24, no. 1, pp. 55-67, 2005.
        [http://dx.doi.org/10.1016/S0278-6125(05)80007-X]

[17]   B. Rouwenhorst, B. Reuter, V. Stockrahm, G.J. van Houtum, R.J. Mantel, and W.H.M. Zijm, "Warehouse design and control: framework and literature review". *Eur. J. Oper. Res.,* vol. 122, pp. 515-553, 2000.
        [http://dx.doi.org/10.1016/S0377-2217(99)00020-X]

[18]   J.X. Gu, M. Goetschalckx, and L.F. McGinnis, "Research on warehouse operation: a comprehensive review", *Eur. J. Oper. Res.,* vol. 177, no. 1, pp. 1-21, 2007
        [http://dx.doi.org/10.1016/j.ejor.2006.02.025]

[19]   T.P. Stank, D.S. Rogers, and P.J. Daugherty, "Benchmarking: applications by third party warehousing firms", *Logistics and Transportation Review,* vol. 30, no. 1, pp. 55-72, 1994.

[20]   M.A. Cohen, Y.S. Zheng, and V. Agrawal, "Service parts logistics: a benchmarking analysis", *IIE Trans.,* vol. 29, pp. 627-639, 1997.
[http://dx.doi.org/10.1080/07408179708966373]

[21]   S.T. Hackman, E.H. Frazelle, P.M. Griffin, S.O. Griffin, and D.A. Vlasta, "Benchmarking warehousing and distribution operations: an input-output approach", *J. Prod. Anal.,* vol. 16, pp. 79-100, 2001.
[http://dx.doi.org/10.1023/A:1011155320454]

[22]   T.R. Collins, M.D. Rossetti, H.L. Nachtmann, and J.R. Oldham, "The use of multi-attribute utility theory to determine the overall best-in-class performer in a benchmarking study", *Benchmarking (Bradf.),* vol. 13, no. 4, pp. 431-446, 2006.
[http://dx.doi.org/10.1108/14635770610676281]

[23]   M.B.M. De Koster, and P.M.J. Warffemius, "American, Asian and third-party International warehouse operations in Europe: A performance comparison", *Int. J. Oper. Prod. Manage.,* vol. 25, no. 8, pp. 762-780, 2005.
[http://dx.doi.org/10.1108/01443570510608592]

[24]   M.B.M. De Koster, and B.M. Balk, "Benchmarking and monitoring international warehouse operations in Europe", *Prod. Oper. Manag.,* vol. 17, no. 2, pp. 175-183, 2008.
[http://dx.doi.org/10.3401/poms.1080.0013]

[25]   D. Dunne, and R. Martin, "Design thinking and how it will change management education", In: *Academy of Management Learning and Education* vol. 5. , 2006, pp. 514-523.

[26]   R. Martin, *The Design of Business: Why Design Thinking is the Next Competitive Advantage.* Harvard Business Press: Boston, MA, 2009.

[27]   S.B. Poulsen, and U. Thøgersen, "Embodied design thinking: a phenomenological perspective", *CoDesign,* vol. 7, no. 1, pp. 29-44, 2011.
[http://dx.doi.org/10.1080/15710882.2011.563313]

[28]   R.J.J. Boland, F. Collopy, K. Lyytinen, and Y. Yoo, "Managing as designing: lessons for organization leaders from the design practice of Frank O. Gehry", *Des. Issues,* vol. 24, no. 1, pp. 10-25, 2007.
[http://dx.doi.org/10.1162/desi.2008.24.1.10]

[29]   T. Brown, "Design thinking", *Harv. Bus. Rev.,* vol. 86, no. 6, pp. 84-92, 141, 2008.
[PMID: 18605031]

[30]   R. Shah, and P.T. Ward, "Lean manufacturing: context, practice bundles, and performance", *J. Oper. Manage.,* vol. 21, no. 2, pp. 129-149, 2003.
[http://dx.doi.org/10.1016/S0272-6963(02)00108-0]

[31]   A. Johnson, W.C. Chen, and L.F. McGinnis, "Large-scale internet benchmarking: technology and application in warehousing operations", *Comput. Ind.,* vol. 61, no. 3, pp. 280-286, 2010.
[http://dx.doi.org/10.1016/j.compind.2009.10.006]

[32]   N. Cross, K. Dorst, and N. Roozenburg, Preface.*Research in Design Thinking.,* N. Cross, K. Dorst, N. Roozenburg, Eds., Delft University Press: Delft, 1992.

[33]   R.J. Boland, and F. Collopy, Design matters for management.*Managing as Designing.,* R.J. Boland, F. Collopy, Eds., Stanford University Press: California, 2004.

[34]   H.A. Simon, *The Sciences of the Artificial.* 3rd ed. MIT Press: London, 1996.

[35]   A. Rylander, "Design thinking as knowledge work: epistemological foundations and practical implications", *Des. Manage. J.,* vol. 4, no. 1, pp. 7-19, 2009.

[36]   B. Lawson, and K. Dorst, *Design Expertise.* Elsevier: London, 2009.

[37]   V. Papanek, *Design for the Real World: Human Ecology and Social Change.* Academy Chicago Publishers: Chicago, 1984.

[38]   N. Cross, "Design Thinking: Understanding How Designers Think and Work", In: *Berg, New York, NY*, 2011.
[http://dx.doi.org/10.5040/9781474293884]

[39]   D.C. Krueger, P.M. Mellat, and S. Adams, "Six Sigma implementation: A qualitative case study using grounded theory", In: *Production Planning & Control*, 2014, p. 889
[http://dx.doi.org/10.1080/09537287. 2013.771414.]

[40]   S.B. Merriam, and R.S. Grenier, *Qualitative Research in Practice: Examples for Discussion and Analysis.* Jossey-Bass: Hoboken, NJ, 2019.

[41]   R. Sanchez-Marquez, J.M.A. Guillem, E. Vicens-Salort, and J.J. Vivas, "A systemic methodology for the reduction of complexity of the balanced scorecard in the manufacturing environment", *Cogent Business & Management,* vol. 7, no. 1, p. 1720944, 2020.
[http://dx.doi.org/10.1080/23311975.2020.1720944]

[42]   M.V. Sunder, and S. Mahalingam, "An empirical investigation of implementing Lean Six Sigma in higher education institutions", *Int. J. Qual. Reliab. Manage.,* vol. 35, no. 10, pp. 2157-2180, 2018.
[http://dx.doi.org/10.1108/IJQRM-05-2017-0098]

[43]   H.P. Ingason, and E.R. Jónsdóttir, "The house of competence of the quality manager", *Cogent Business & Management,* 2017.
[http://dx.doi.org/10.1080/23311975.2017.1345050]

[44]   S. Vinodh, S.V. Kumar, and K.E.K. Vimal, "Implementing Lean Sigma in an Indian Rotary Switches Manufacturing Organization", *Production Planning & Control,* vol. 25, no. 4, pp. 288-302, 2014.

[45]   M.V. Sunder, L.S. Ganesh, and R.R. Marathe, "Lean Six Sigma in consumer banking: An empirical inquiry", *Int. J. Qual. Reliab. Manage.,* vol. 36, no. 8, pp. 1345-1369, 2019.
[http://dx.doi.org/10.1108/IJQRM-01-2019-0012]

[46]   A. Bazrkar, S. Iranzadeh, and N.F. Farahmand, "Total quality model for aligning organization strategy, improving performance, and improving customer satisfaction by using an approach based on combination of balanced scorecard and lean six sigma', *Cogent Business & Management,* 2017.
[http://dx.doi.org/10.1080/23311975.2017.1390818]

[47]   H. Julien, Content analysis.*The SAGE Encyclopedia of Qualitative Research Methods.,* L.M. Given, Ed., SAGE Publications: Thousand Oaks, CA, 2008.

[48]   H.F. Hsieh, and S.E. Shannon, "Three approaches to qualitative content analysis", *Qual. Health Res.,* vol. 15, no. 9, pp. 1277-1288, 2005.
[http://dx.doi.org/10.1177/1049732305276687] [PMID: 16204405]

[49]   R.P. Weber, *Basic Content Analysis.* SAGE University Paper: Thousand Oaks, CA, 1990.
[http://dx.doi.org/10.4135/9781412983488]

[50]   L.J. Cronbach, *Educational Measurement.* 2nd ed. American Council on Education: Washington, DC, 1971.

[51]   E.R. Babbie, *The Practice of Social Research.* 10th ed. Thomson/Wadsworth: Belmont, CA, 2004.

[52]   E.P. Michel, *Competitive Advantage: Creating and Sustaining Superior Performance.* Free Press: New York, 1985, pp. 3-52.

[53]   M.D. Sayidmia, "An approach to reduce Manufacturing waste and Improve the Process Cycle Efficiency of a footware Industry by using Lean Six-Sigma Model", *Master of Science in Management of Technology Dissertation,* pp. 1-153, 2016. Institute of Appropriate Technology. Bangladesh University of Engineering and Technology; Available at, http://lib.buet.ac.bd:8080/xmlui/bitstream/handle/123456789/4511/Full%20Thesis.pdf?sequence=1&isAllowed=y. [Accessed 10th January, 2021].

[54]   C.K. Swank, "The lean service machine", *Harv. Bus. Rev.,* vol. 81, no. 10, pp. 123-130, 2003.
[PMID: 14521103]

[55]   D.T. Jones, P. Hines, and N. Rich, "Lean logistics", *International Journal of Physical Distribution & Logistics Management,* vol. 27, p. 153, 1997.
[http://dx.doi.org/10.1108/09600039710170557]

[56]   H.D. Wan, and F. Chen, "A leanness measure of manufacturing systems for quantifying impacts of lean initiatives", *Int. J. Prod. Res.,* vol. 46, no. 23, pp. 6567-6584, 2008.
[http://dx.doi.org/10.1080/00207540802230058]

[57]   J.R. Hackman, and G.R. Oldham, *Work Redesign.* Addison-Wesley Publishing Company: Reading, 1980.

[58]   M. Vidal, "Lean production, worker empowerment, and employee satisfaction: a qualitative analysis and critique", *Critical Sociology,* vol. 33, pp. 247-278, 2007.

[59]   S. Bhasin, and P. Burcher, "Lean viewed as a philosophy", *J. Manuf. Tech. Manag.,* vol. 17, no. 1, pp. 56-72, 2006.
[http://dx.doi.org/10.1108/17410380610639506]

[60]   B.A. Henderson, J.L. Larco, and S. Martin, *Lean transformation: how to change your business into a lean enterprise.* Oaklea Press: Richmond, VA, 1999.

CHAPTER 10

A Markovian Analysis of Industrial Accident Data in a Nigerian Manufacturing Company

Kazeem Aderemi Bello[1,*], **Olatunde A. Oyelaran**[1], **Ilesanmi Afolabi Daniyan**[2] and **Osarobo Osamede Ogbeide**[3]

[1] *Department of Mechanical Engineering, Federal University, Oye-Ekiti, Nigeria*

[2] *Department of Industrial Engineering, Tshwane University of Technology, Pretoria 0001, South Africa*

[3] *Production Engineering Department, University of Benin, Benin City, Nigeria*

Abstract: Despite the effort of manufacturing company owners in Nigeria to curtail industrial accident occurrence, it still remains a daunting challenge. This study seeks to predict the drift of industrial accidents in manufacturing companies in Delta State, Nigeria in order to ensure the health and safety of staff in the workplace. Markov Chain (MC) model was used to analyse eleven year industrial accident data obtained from a primary source: Health, Safety, and Environment (HSE) department of a manufacturing company in the Delta state of Nigeria. The data was summarised as integrated safety data to give a better picture of the organisation safety culture. The data was analysed and found to have absorbing chain tendencies. It also possesses a note of stochastic regularity, which fits into an MC model. Accident data in the company were classified into five states, namely fatality, loss time accident, medical aid accident, first aid incident, and near-miss. The result from the study reveals that fatality and loss time accidents were absorbing states, while medical aid accidents, first aid incidents, and near-miss were found to be non-absorbing states. A worker who commits an error in the form of near miss 1000 times stands a chance of fatality 77 times out of this total time of work error. This study will serve as a guide to manufacturing company stakeholders on the need to create safety awareness among the workforce.

Keywords: Absorbing, Accident, Markov chain, Non-absorbing, Transition Matrix.

INTRODUCTION

MC is a random process widely used in modelling a wide range of situations. The two assumptions common with Markov are its ability to only depend on the current state irrespective of the event that preceded it, otherwise called the memo-

---
* **Corresponding Author Kazeem Aderemi Bello:** Department of Mechanical Engineering, Federal University, Oye-Ekiti, Nigeria; Tel: +2348036386760; E-mail: kazeem.bello@fuoye.edu.ng

Ilesanmi Afolabi Daniyan (Ed.)

memoryless of a Markov Property and time homogeneous property [1]. The evolution of a random process of a system over time can be modeled as an MC. There are two basic assumptions. The first is "Memory less" of a Markov property, as observed by Igboanugo *et al.* [2], who defined Markov as a "series of states of a dynamic system that has the Markov property". Further, MC has the ability to change from one state to another without keeping records on all the sequences except the preceded event. The second assumption is the time-homogeneous nature of transition probabilities that are not dependant on time. These requirements are not often fulfilled completely.

A Russian Mathematician, Andrei Andreevich Markov (1856-1922), who lived in the early 20th century, studied the theory of stochastic process in 1907, which metamorphoses into what is referred to as the Markov Chain. Modelling practical methods to ascertain the trend of the pattern was the first application of Markov Chain. Random physical properties can be modelled using Markov chains. Markov Chain is a stochastic process. This refers to a system whose behaviour is intrinsically non- deterministic, occurring irregularly and non-intermittent.

Markov chain theory has wide or extensive application. In a manpower system, the Markov chain can be used to predict, describe, as well as control personnel mobility through the different states. The states could refer to personnel records of recruitment, retirement, training, promotion, as well as staff stock [3].

Markovian planning in a soft drink industry in Lagos, Nigeria [4] focused on the short-term and long-term challenges with regard to long-run policies as it applied to absorbing and non-absorbing states (retirement, promotion, wastage, and recruitment). Transition probabilities matrix were generated and computed.

Moreover, a sequence of trials in which the outcome of any particular trial depends only upon the outcome of the immediately preceding trial is known as the Markov chain or Markov process.

The Factories' act, the Workmen's compensation act, and the labour safety, health, and welfare bill [5 - 8] are important documents aimed at protecting the health and safety of the Nigerian workers. Amponsah-Tawiah and Dartey-Baah [9] narrated the issues of concern as related to occupational health in the workplace. Hazards elimination in the workplace is the best way to guarantee the safety of workers [10]. Other researchers and many more emphasized the significance of zero accidents in the workplace [10 - 13]. The Markov chain has been widely applied in market research, advertising, sales forecasting, industrial accident analysis, and other management problems.

METHODOLOGY

An eleven-year accident data were collated in the company as recorded in Table **1**, while the summary of the data is presented in Table **2**. The incidents were collated annually for five different classifications, namely: fatality (s_1), loss time accident (s_2), medical aid accident (s_3), first aid incident (s_4), and near-miss (s_5).

Table 1. Yearly workers' safety records in fibre cement roofing sheet manufacturing company for eleven years (2005-2015).

					Years							Total
States	2005	2006	2007	2008	2009	2010	2011	2012	2013	2014	2015	
Fatality (FAT)	0	0	0	0	0	0	0	0	1	0	0	1
Loss Time Accident (LTA)	8	6	4	2	6	5	1	0	0	1	3	36
Medical Aid Accident (MAA)	18	15	12	6	15	12	3	5	3	4	2	95
First Aid Incident (FAI)	25	17	19	12	8	22	9	6	7	5	6	136
Near Miss (NM)	40	35	32	25	34	41	14	8	10	12	8	259
TOTAL	91	73	67	45	63	80	27	19	21	22	19	527

Table 2. An eleven-year HSE record summary.

States	No of occurrence	Proportion (%)
Fatality (FAT)	1	0.19%
Loss Time Accident (LTA)	36	6.83%
Medical Aid Accident (MAA)	95	18.03%
First Aid Incident (FAI)	136	25.81%
Near Miss (NM)	259	49.15%
Total	527	100%

Fatality (FAT): Is a work-related accident that involves an employee or a contractor working on site (or offsite), resulting in loss of life regardless of the length of time between the date of the accident and the date the person dies.

Loss Time Accident (LTA): Is an occurrence causing an injury resulting in the absence of the employee for one or more calendar days, counting from the day after the injury occurs to the day before the individual employee returns to normal or modified work.

Medical Aid Accident (MAA): Is an occurrence causing an injury that requires treatment by a trained health care professional (nurse, doctor, physiotherapist, *etc.*) and where the worker comes back to work the next day/shift.

First Aid Incident (FAI): Is a non-lost time work-related occurrence with a minor injury that requires a one-time treatment or minor aches and pains, scratches, cuts, burns as so forth, that does not require (or should not have required) medical care by a trained health care professional.

Near- Miss: An incident or situation which, in the opinion of the observer, could have resulted in personal injury or ill health if circumstances were different. This includes unsafe acts, unsafe behaviours, and unsafe situations.

Fundamental Matrix Model Development

The from-to-matrix associated with the computed Transition Probability Matrix (TPM) OS is represented in Table **3**.

Table 3. From-to-matrix.

		FAT	LTA	MAA	FAI	NM
Absorbing	FAT	P_{11}	P_{12}	P_{13}	P_{14}	P_{15}
	LTA	P_{21}	P_{22}	P_{23}	P_{24}	P_{25}
	MAA	P_{31}	P_{32}	P_{33}	P_{34}	P_{35}
Non-Absorbing	FAI	P_{41}	P_{42}	P_{43}	P_{44}	P_{45}
	NM	P_{51}	P_{52}	P_{53}	P_{54}	P_{55}

Modelling with Markov Chain

A Markov Chain can model a situation that is seen as having reoccurrence ability (trials) among various classifications. Labeling the states $S_1, S_2, S_3\ldots\ldots\ldots\ldots S_n$ and for each pair (S_i, S_j), a fixed proportion P_{ij} of occurrence located at S_i is moved to S_j during each step or movement. Thus the scenario can be represented in a transition probability matrix as indicated in eq (2).

Moreover, every occurrence is seen as having the chance to eventually reach an absorbing state on account of which the system is being called an absorbing chain.

In this regard, for a partitioned matrix in the standard form shown in equations (1) to (4)

$$T = \begin{bmatrix} I & O \\ \hline R & Q \end{bmatrix} \tag{1}$$

$$T^2 = \begin{bmatrix} I & O \\ \hline R & Q \end{bmatrix} * \begin{bmatrix} I & O \\ \hline R & Q \end{bmatrix}$$

$$T^2 = \begin{bmatrix} I & O \\ \hline (I+Q)R & Q^2 \end{bmatrix} \tag{2}$$

$$T^3 = \begin{bmatrix} I & O \\ \hline R & Q \end{bmatrix} * \begin{bmatrix} I & O \\ \hline R & Q \end{bmatrix} * \begin{bmatrix} I & O \\ \hline R & Q \end{bmatrix}$$

$$T^3 = \begin{bmatrix} I & O \\ \hline (I+Q+Q^2)R & Q^3 \end{bmatrix}$$

Also

$$T^4 = \begin{bmatrix} I & O \\ \hline R & Q \end{bmatrix} * \begin{bmatrix} I & O \\ \hline R & Q \end{bmatrix} * \begin{bmatrix} I & O \\ \hline R & Q \end{bmatrix} * \begin{bmatrix} I & O \\ \hline R & Q \end{bmatrix} \tag{3}$$

$$T^4 = \begin{bmatrix} I & O \\ \hline (I+Q+Q^2+Q^3)R & Q^4 \end{bmatrix}$$

Thus in general,

$$T^n = \begin{bmatrix} I & O \\ \hline (I+Q+Q^2+...+Q^{n-1})R & Q^n \end{bmatrix} \tag{4}$$

Clearly

(I-Q) (I+Q+Q²+...+Qⁿ⁻¹) R=1-Qⁿ

$$\therefore T^n = \begin{array}{c} \\ abs \\ \\ nonabs \end{array} \overset{\begin{array}{cc} abs & nonabs \end{array}}{\left[\begin{array}{c|c} I & O \\ \hline (I-Q)R & Q^n \end{array}\right]} \tag{5}$$

The procedure adopted include determination of transition probability matrix, T, as shown in equation (1).

$$T = \begin{bmatrix} P_{11} & P_{12} & P_{13} & .... & P_{11K} \\ P_{21} & P_{22} & P_{23} & ... & P_{21K} \\ ... & ... & ... & ... & ... \\ P_{K1} & P_{K2} & P_{K3} & ... & P_{KK} \end{bmatrix} \tag{6}$$

It is assumed that T is standard.

$$\sum P_{ij} = 1 \forall i$$

$$\sum P_{ij} \geq 0, \forall i \ and \ j$$

Partitioned matrix, *i.e.*, the fundamental matrix, was developed, and interpretations were made using Markov theories.

$$T = \begin{array}{c} \\ S_1 \\ S_2 \\ ... \\ S_K \end{array} \overset{\begin{array}{cccc} S_1 & S_2 & ... & S_K \end{array}}{\begin{bmatrix} P_{11} & P_{12} & ... & S_K \\ P_{21} & P_{22} & ... & P_{1K} \\ ... & ... & ... & ... \\ P_{K1} & P_{K2} & ... & P_{KK} \end{bmatrix}} \tag{7}$$

$T_{ij}=P_{ij}$

Theorem 1

An object from an initial distribution d_o in a Markov chain can move through many transitions to attain a long-run distribution, d_n and hence achieve a stable transition distribution probability matrix $\bar{T}_L$ with stabilized distribution as shown in equation (8)

$$d_n = d_o T^n \tag{8}$$

Where $\bar{T}_L$ stable transition distribution matrix.

Theorem 2

Thus, if as the number of transitions increases $(n\uparrow)$, the transition matrix T approaches a matrix $\bar{T}$; and then for each choice of an initial distribution d_o, the subsequent distributions can be expressed like so:

$d_n = d_o T^n$ stabilize to the long run distribution given by

$$\bar{d} = d_o \bar{T} \tag{9}$$

where $\bar{T}$ is the stabilized transition matrix.

Moreover, each $\bar{d}$ arising in this fashion is stationary.

Theorem 3

Let T be the transition matrix of a regular Markov chain, then the power of T approach a matrix $\bar{T}$ with long-run distribution, all of whose rows are identical and all entries positive.

There remains the practical problem of actually computing the unique long-run distribution $\bar{d}$ associated with $\bar{T}$ of regular Markov chain.

d = dT, for stationary d and regular T

= dI = dT

= dI − dT = 0

and d (I - T) = 0

Take the transpose as shown in (5)

$$\Rightarrow \theta(I-T)^t d^t = 0^t \Leftarrow AX = 0 \tag{10}$$

Equation (10) is a homogeneous system.

This leads to a theorem

Theorem 4

Every Markov Chain with k states has at least one long-run distribution. Moreover, each long-run distribution is the transpose of a solution to:

AX = 0

Whose entries sum to one here

A = (I – T)^t,

T = k-by-k transition matrix, and both X and 0 are k-by-I column vectors.

Theorem 5

Consider absorbing the Markov chain with a transition matrix.

$$\bar{T} = \begin{matrix} Absorbing \\ \\ Nonabsorbing \end{matrix} \begin{bmatrix} I & - & O \\ - & - & - \\ (I-Q)^{-1}R & - & O \end{bmatrix}$$

The ith row of the submatrix B = (I-Q)$^{-1}$R

The Transition Probability Matrix (TPM) diagram is presented in Fig. (**1**).

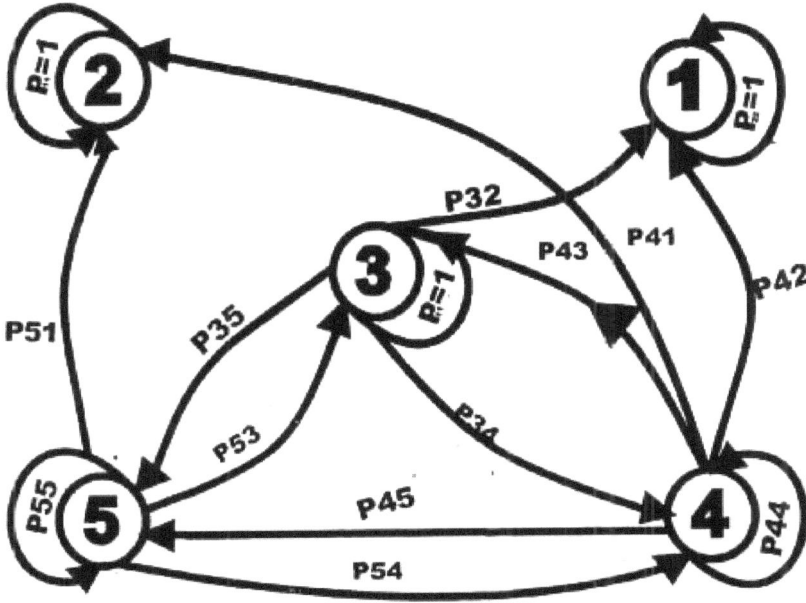

Fig. (1). Transition Probability Matrix Diagram.

Statistical Computations

From Fig. **(1)**, it was noted that absorbing state transition probabilities are heuristically determined.

P_{11}- P_{15} and $P_{21} - P_{25}$ were calculated based on Heuristics and probability arguments.

$$P_{11}=1 \; P_{12}=0 \; P_{13}=0 \; P_{14}=0 \; P_{15}=0$$
$$P_{21}=0 \; P_{22}=1 \; P_{23}=0 \; P_{24}=0 \; P_{25}=0$$

The values of P_{31} - P_{35}, $P_{41} - P_{45}$, and $P_{51} - P_{55}$ were calculated based on the transition probability of elements in the various states (absorbing and non-absorbing states).

$P_{31} = \dfrac{1}{493} = 0.0020$, $P_{32} = \dfrac{0}{493} = 0$, $P_{33} = \dfrac{95}{493} = 0.1927$, $P_{34} = \dfrac{136}{493} = 0.2759$, $P_{35} = \dfrac{259}{493} =$ 0.5254.

$P_{41} = \dfrac{1}{529} = 0.0019$, $P_{42} = \dfrac{36}{529} = 0.0681$, $P_{43} = \dfrac{95}{529} = 0.1796$, $P_{44} = \dfrac{136}{529} = 0.2571$, $P_{45} = \dfrac{259}{529} =$ 0.4896

$P_{51} = 0$, $P_{52} = \dfrac{36}{528} = 0.0682$, $P_{53} = \dfrac{95}{528} = 0.1799$, $P_{54} = \dfrac{136}{528} = 0.2578$, $P_{55} = \dfrac{259}{528} = 0.4905$

Table **4** presents the results obtained from the numerical computation of TPM.

Table 4. Numerical computation of TPM.

		FAT	LTA	MAA	FAI	NM
Absorbing	FAT	1	0	0	0	0
	LTA	0	1	0	0	0
	MAA	0.0020	0	0.1927	0.2759	0.5254
Non-Absorbing	FAI	0.0019	0.0681	0.1796	0.2571	0.4896
	NM	0	0.0682	0.1799	0.2578	0.4905

The values are expressed in compact form as follow:

$$T = \begin{bmatrix} 1 & 0 & 0 & 0 & 0 \\ 0 & 1 & 0 & 0 & 0 \\ 0.0020 & 0 & 0.1927 & 0.2759 & 0.5254 \\ 0.0019 & 0.0681 & 0.1796 & 0.2571 & 0.4896 \\ 0 & 0.0682 & 0.1799 & 0.2578 & 0.4905 \end{bmatrix}$$

$$\left[\begin{array}{c|c} I & O \\ \hline R & Q \end{array} \right]$$

Where:

$$I = \begin{bmatrix} 1 & 0 \\ 0 & 1 \end{bmatrix}$$

$$R = \begin{bmatrix} 0.0020 & 0 \\ 0.0019 & 0.0681 \\ 0 & 0.0682 \end{bmatrix} \quad \text{R stands for initial distribution}$$

$$Q = \begin{bmatrix} 0.1927 & 0.2759 & 0.5254 \\ 0.1796 & 0.2571 & 0.4896 \\ 0.1799 & 0.2578 & 0.4905 \end{bmatrix} \text{Where Q is with transitions from non-transient to transient}$$

$$I - Q = \begin{bmatrix} 1 & 0 & 0 \\ 0 & 1 & 0 \\ 0 & 0 & 1 \end{bmatrix} - \begin{bmatrix} 0.1927 & 0.2759 & 0.5254 \\ 0.1796 & 0.2571 & 0.4896 \\ 0.1799 & 0.2578 & 0.4905 \end{bmatrix}$$

The identity matrix was regularized as shown above for ease of computations, which were carried out using MATLAB software.

$$I = Q = \begin{bmatrix} 0.8073 & -0.2759 & -0.5254 \\ -0.1796 & 0.7429 & -0.4896 \\ -0.1799 & -0.2578 & 0.5095 \end{bmatrix}.$$

$$N = (1-Q)^{-1} = \begin{bmatrix} 4.2344 & 4.6327 & 8.8184 \\ 3.0142 & 5.3172 & 8.2178 \\ 3.0203 & 4.3262 & 9.2345 \end{bmatrix}$$

Where $(I-Q)^{-1}$ refer to mean habitation of the accident within the non-absorbing matrix. This is also referred to as the fundamental matrix

$$NR = (1-Q)^{-1} R = B = \begin{bmatrix} 4.2344 & 4.6327 & 8.8184 \\ 3.0142 & 5.3172 & 8.2178 \\ 3.0203 & 4.3262 & 9.2345 \end{bmatrix} \times \begin{bmatrix} 0.0020 & 0 \\ 0.0019 & 0.0681 \\ 0 & 0.0682 \end{bmatrix}$$

$$B = \begin{bmatrix} 0.0773 & 0.9169 \\ 0.0161 & 0.9226 \\ 0.0943 & 0.9244 \end{bmatrix}$$

RESULTS AND DISCUSSION

The transition probability matrix was developed from the five states industrial accident. Accordingly, the values were subsequently fed into MATLAB 2007b (7.5) software with a guided formulation that provided the following output:

i. The long run distribution of workers before absorbing states.
ii. The total number movement of workers within the non-absorbing states, and
iii. Transition of workers from non-absorbing to absorbing states.

The interpretations of the results were based on the mathematical analysis and presented accordingly. The fundamental matrix provides information about the mean number of habituations, which objects undergo within the non-absorbing states before being finally absorbed. The matrix is shown below:

$$N = (1 - Q)^{-1} = \begin{bmatrix} 4.2344 & 4.6327 & 8.8184 \\ 3.0142 & 5.3172 & 8.2178 \\ 3.0203 & 4.3262 & 9.2345 \end{bmatrix}$$

Based on the above matrix, the following interpretations are rendered:

$3 \rightarrow 5$ with entry value 8.8184. This suggests that a worker commits up to nine near misses before committing medical aid accident.

$4 \rightarrow 3$ with an entry value of 4.2344. The interpretation is that a worker only needs to commit 4 mistakes to move from first aid incident to medical aids accident.

$4 \rightarrow 4$ has an entry of 5.3172. This suggests that a worker needs to undergo a minimum of five times safety awareness briefs in a week to work safely.

$4 \rightarrow 5$ with an entry value of 8.2178. This result suggests that a worker who commits eight mistakes may likely transit to the next states.

$5 \rightarrow 3$ with entry value of 3.0203.This suggests that a worker commits a minimum of three near misses in a year.

$5 \rightarrow 5$ with the entry value 9.2345. This suggests that a worker needs to be trained for a minimum of nine days on safety before being engaged to work in the factory.

(II) $\rightarrow$ The total number of movements of workers within the non-absorbing states.

(III) B – Matrix: Long run transition within the state.

The movement is explained in matrix B.

$$\begin{bmatrix} 17.6856 \\ 16.5492 \\ 16.5810 \end{bmatrix}$$

The matrix above revealed that generally, all workers, irrespective of the starting state, undergo 17 transitions among the three none absorbing situations before being trapped in any of the absorbing situations.

Here we evaluate the probability that the worker is absorbed in S_1 or S_2 given that he starts in S_3, S_4, and S_5.

$$B = \begin{bmatrix} 0.0773 & 0.9169 \\ 0.0161 & 0.9226 \\ 0.943 & 0.9244 \end{bmatrix}$$

The probability of committing loss time accident if the worker starts in S_3 (*i.e.*, be absorbed in S_3 is 0.9169). The probability of committing a loss time accident if the worker starts in S_4 is 0.9226, while the probability of committing a lost time accident if the worker starts in S_5 is 0.9244. However, the probability of being killed if the worker starts in S_3 is 0.0773; the probability of being killed if the worker starts from S_4 is 0.0161, while the probability of being killed if the worker starts from S_5 is 0.943.

The Markov Chain model employed has shown that some quantities relating to absorbing chains have utilities in the analysis of industrial accident data, among others. The quantity called the "fundamental matrix" given by

N= $[I - Q]^{-1}$ is central to the analysis with Markov Chains. The fundamental matrix is derived from the canonical form involving partition into four portions comprising I, O, R, and Q. Procedures have been developed in this study for determining the following:

1. The number of times a worker habituated within either of the transient states $\{S_3, S_4, S_5\}$ before entering the axis of absorbing state.
2. It also enabled the determination of the average resident time a worker spent working in the organisation before being involved in an accident that may lead to either incapacitation or death.

The accident rate in the organisation studied is relatively low (less than 1% of the workers die). It is a reflection of the level of safety training and pre-occupation orientation in which the workers are fully involved. The computed standard deviation associated with the habituation times before being involved in an accident appears very close, thus implying that resident times in a particular state could be much lower than the computed averages. For instance, when a worker commits a work error in the form of near miss 1000 times, such a worker stands a chance of fatality 77 times out of this total time of work error.

CONCLUSION

The MC model adopted for this study has been successful in analysing industrial accidents. The study reveals that industrial accident victims habituate several times before entering the absorbing states, unlike the road traffic accident victims. Despite the safety warning signs and safety training given to all levels of staff, noteworthy that some workers still fail to observe safety rules. The management, however, needs to demonstrate enough commitment to enforce safety compliance among the workforce. It is shown in the results that a worker who commits an error in the form of near miss 1000 times stands a chance of fatality 77 times out of this total time of work error. This result is commendable but still far from the management objective of zero accidents.

CONSENT FOR PUBLICATION

Not applicable.

CONFLICT OF INTEREST

The authors declare no conflict of interest, financial or otherwise.

ACKNOWLEDGEMENTS

Declared none.

REFERENCES

[1]   S. Ross, *Stochastic Process.* 2nd ed. John Wiley: New York, 1996.

[2]   A.C. Igboanugo, and M.K. Onifade, "Markov Chain analysis of manpower data of a Nigerian University", *Journal of Innovative Research in Engineering and Science,* vol. 2, no. 2, pp. 107-123, 2011.

[3]   K. Setlhare, "Modeling of an intermittently busy manpower system", *Conference Gaborone, Botswana,* 2007

[4]   A.C. Igboanugo, and O.R. Edokpia, "A Markovian Study of manpower planning in the soft-drink industry in Nigeria", *Journal of Research in Engineering,* vol. 3, pp. 24-30, 2008.

[5]   A.C. Munthali, and J. Milner, *Recording and Notification of Occupational accident and diseases in Malawi.* International Labour Organisation: Lusaka, 2012.

[6]   "Abstract of the Act in form prescribed by the Federal ministry of Labour and Productivity. ", In: *Director of Factories, Federal ministry of Labour and Productivity, Federal Secretariat, Private Mail Bag 04 Garki-Abuja* Nigeria, 1990.

[7]   H.A. Gezairy, "Occupational health: A manual for primary health care workers. World Health Organization", *Regional Office for the Eastern Mediterranean Cairo,* 2001. WHO-EM/OCH/85/E/L Distribution Limited

[9]   K. Amponsah-Tawiah, and K. Dartey-Baah, "Occupational health and safety: key issues and concerns in Ghana", *Int. J. Bus. Soc. Sci.,* vol. 2, no. 14, pp. 199-126, 2014.

[10]   K.A. Bello, and A.C. Igboanugo, "Hazard identification and control in workplace: A case study of fibre cement roofing sheet manufacturing company", *International Journal of Scientific and Engineering Research,* vol. 8, pp. 1801-1813, 2017.

[11]   *Prevention: A global Strategy Promoting Safety and Health at Work. The I.L.O Report for World day of Safety and Health at Work.* International Labour Office Geneva, 2005.

[12]   J.C. Dopkeen, *Stress in Workplace: A policy Synthesis or its Dimensions and Prevalence.* Centre for Employee Health Studies, Health Administration: Chicago, Illinois, 2014.

[13]   M.R. Kumar, K.N. Karthick, T. Dheenathayalan, and K. Visagavel, "Exposure hazard analysis in cement fibre sheet manufacturing industry", *Int. J. Res. Eng. Technol,* vol. 3, no. 11, pp. 76-80, 2014.

CHAPTER 11

Assessment of Manufacturing Productivity in Fibre Cement Roofing Sheet Production Company

Kazeem Aderemi Bello[1,*], Olatunde A. Oyelaran[1], Ilesanmi Afolabi Daniyan[2] and Osarobo Ogbeide[3]

[1] *Department of Mechanical Engineering, Federal University, Oye-Ekiti, Nigeria*

[2] *Department of Industrial Engineering, Tshwane University of Technology, Pretoria 0001, South Africa*

[3] *Production Engineering Department, University of Benin, Benin City, Nigeria*

Abstract: The absence of adequate training, technical know-how, and process control skills may be associated with high product rejects and high rate of industrial accident occurrence in the course of fibre cement roofing sheet manufacture. The overall research strategy of this study entailed the use of Kendall Coefficient of Concordance (KCC) and Principal component analysis (PCA) to investigate the identified factors that influence the production of fibre cement roofing sheets. Statistical Process Control (SPC) tools were used to analyse customers' complaints and preferences. The Markov Chain (MC) model was also used to analyse industrial accident occurrence trends. Also, multiple linear regression analysis was conducted to predict the pattern of productivity of fibre cement roofing sheet products.

Our results show that Kendall coefficient of concordance employed ranked the sixty-one variables of fibre cement roofing sheet in descending order of importance. Furthermore, SPC results on customers' complaints revealed sticking sheets, broken edges, and weak lines as the main cause of product defects, while the inadequate maintenance on production machines was not identified as the root cause. Due to the implementation of Quality Management Committee (QMC) suggestions, the product process capability was boosted to 1.19. Product density of 600 observations was considered, and all observations fell within the R and X charts control limit. This research will provide guidelines for roofing sheet producers to manufacture quality fibre cement roofing sheet that is suitable for customers' needs.

Keywords: Accident, Customer satisfaction, Fibre cement roofing sheet, Productivity, Quality control.

* **Corresponding Author Kazeem Aderemi Bello:**Department of Mechanical Engineering, Federal University, Oye-Ekiti, Nigeria; Tel: +2348036386760; E-mail: kazeem.bello@fuoye.edu.ng

Ilesanmi Afolabi Daniyan (Ed.)

INTRODUCTION

In terms of consuming services and goods, humanity is geared to choose the most cost effective options. Therefore, manufacturers who desire to stand the test of time must work hard to produce good quality products at affordable prices, which can compete favourably with alternative products in the marketplace. Hence, a conscious effort should be made to advance the manufacturing of goods in Nigeria to meet internationally acceptable standards in all manufacturing sectors, most especially in the roofing and building material manufacturing sector. The reasons for poor patronage of locally manufactured goods could be linked to poor quality, lack of innovation and technological input, lack of research and development, poor customer relation, *etc.* In order to win customer loyalty to homemade goods, a paradigm shift in the Nigerian manufacturing industry is required to match customer satisfaction and demands. There is no doubt about the fact that if the products are not adequately patronised, the manufacturer business will close by natural principle over time. Therefore, there is a need to continually advance the quality of manufacturing products to meet customer requirements and satisfaction in order to sustain the manufacturing companies and remain in business.

Etex World [1] narrated how fibre cement was invented in 1890 by the Austrian Industrialist Ludwing Hatschet. Again, the report pointed out how the concept of fibre cement was originated unexpectedly by mixing cement with asbestos fibres and water, and the resulting material had very interesting properties. It was durable and strong, yet light in weight. It was fire-proof and resistant to humidity and frost. It could also be formed into many different shapes, such as roofing sheets, flat boards for facades and interior finishing, pipes, flower pots, *etc.* Later on, asbestos fibres were replaced by other types of fibres such as cellulose and polyvinyl alcohol (PVA).

It has been noted that the implementation of local content policy in fibre cement roofing sheet manufacturing remains unachievable despite the efforts of government to do so [2]. Igboanugo *et al.* [2] and Balonga [3] noted further that the inclusion of local content policy would enable manufacturers to reduce the cost of production and help promote Nigeria's economy. To improve the strength of the Modulus of Rupture (MOR) of fibre cement, Goezeglanczyk *et al.* [4] utilised two separate methods, such as cause and effect matrix and failure mode and effect analysis to reduce the number of factors to be studied further. The result showed a process capability index (PCI) improvement. Some researches on fibre cement roofing production include eucalyptus pulp refining effects and durability of fibre cement composites [5], light-weight fibre cement cladding elements [6], fibre cement recycling from the grave back to the cradle, flexural performance of agro-waste cement composite, optimal use of flocculent on the manufacturer of fibre-cement materials [7, 8], *etc.* Frazao and Fernandes [9] carried out a comparative analysis of the life cycle of ATFibre-cement and NT fibre-cement. The objective of the study is to compare the life cycle of two types of roofing sheets, namely ATFibre-cement and NT fibre-cement, in terms of eco-efficiency using a comparative Life Cycle Assessment (LCA) and Life Cycle Costs (LCC) of each type of sheet. The results obtained indicate that the ATFibre-cement sheets are more economical and ecologically efficient than the NT fibre-cement. Previous studies indicate that the Asbestos-Cement (AC) roof sheets are liable to surface corrosion and, as such, may release toxic fumes, which are carcinogenic [10, 11]. Hence, the use of the asbestos-cement poses a significant health risk. Due to the health risk associated with the use of asbestos-cement, Lin [12] advocated the transformation to fibre cement from asbestos cement. However, Bello *et al.* [13] employed the survey approach to uncover some of the challenges of fibre cement roofing sheet manufacturing from the customers' perspectives. Some of the identified challenges include weak lines, broken edges, sticking sheets, cracks, delimitation, wet sheets, broken pieces, length defects, and rough edges.

This study seeks to survey the key variables that affect the quality of roofing sheets, ascertain their individual and collective roles in quality control, and employ Statistical Process Control (SPC) for quality control. The purpose of this research, therefore, is to sensitise manufacturing firms on the need to adopt good engineering practices in the manufacturing and maintenance of production facilities.

The overall research strategy of this study entailed the use of Kendall Coefficient of Concordance (KCC) and Principal component Analysis (PCA) to investigate the identified factors that influence the production of fibre cement roofing sheets. SPC tools were used to analyse customers' complaints and preferences.

METHODOLOGY

The focus of this study was to survey a gamut of variables that affect the operations of fibre cement roofing sheet manufacturing and the inter correlation among the variables. Also, this study seeks to investigate customers' complaints on dysfunctional aspects of fibre cement roofing products quality that satisfy customers' needs using statistical tools to study the variables that affect or impact fibre cement manufacturing. The study employs the Statistical Process Control (SPC) for quality control. The use of the SPC as statistical technique for process monitoring and control has been demonstrated [14 - 16]. The choice of the **PCA** technique stems from the fact that it is suitable for representing a multivariate data table as a smaller set of variables with summary indices. This will allow precise observation of trends, clusters, outliers or jumps, clusters and outliers [17, 18]. This will enable the determination of the relationships between or among the variables.

Markov chain was employed to analyse industrial accidents in the workplace to assess the level of safety compliance and to ensure the health and safety of the workforce. The research design used in this study includes the administration of questionnaires and scaled on Resis Likert's 5-points attitudinal range. The questions were given to thirteen selected judges who ranked the sixty-one scale items according to the order of relevance. The consistency in the judges' ability to rank the variables is determined by KCC. Chi-square (χ^2) statistic was implored to rate how capable are the judges while ranking the variables.

Again, questions were crafted with the 61 scale items and administered to respondents. The respondents' results were gathered and analysed using StatistiXL software. KCC was used to analyse the variables according to their merit order of importance, while the PCA was used for factor reduction. The results such as scree plot, eigenvalue and factor loadings, and descriptive statistics, communalities, varimax rotated factors loadings, among others, were obtained and then interpreted.

Furthermore, multiple linear regressions were used to forecast the output of quality of fibre sheet production from the data obtained for 128 consecutive months (January 2006 to August 2016). Finally, investigation, analysis, and recommendations on how to improve the quality of fibre cement roofing sheets that meet customer needs, requirements, and satisfaction at affordable prices were suggested.

Data Source and Size

Thumbnail sketches of the source of data are presented hereunder:

 i. Relevant information from the past studies, *e.g.,* through journals, relevant reports, dissertations, *etc.*
 ii. Questionnaire, which forms the basis for the survey aspect of the research, was carefully crafted based on information obtained from (i) above and administered to knowledgeable people of not less than 15 years cognate experience in fibre cement roofing sheet manufacturing operation.
 iii. Records of 36-month customers' complaints record were collated with available records and then analysed.
 iv. Furthermore, multiple linear regression was used to forecast the output of

quality of fibre sheet production from the FMS data obtained that for 128 consecutive months (January 2006 to August 2016).

Models Employed

i. KCC and PCA were used to investigate the roles of and interplay among the identified factors that affect the production of fibre cement roofing sheets.

ii. Statistical tools were adopted to analyse customers' complaints and preferences. Minitab and Quality Management Committee (QMC) software package employed facilitated the analysis.

iii. Markov Chain model was used to analyse the industrial accident with the intention of discerning trends in industrial accident occurrences in order to ensure the health and safety of staff in the workplace.

iv. Regression analysis was conducted with the purpose of predicting the pattern of productivity of fibre cement roofing sheet products.

Kendall's Coefficient of Concordance Analysis

The 61 variables were referred to 13 judges who are well-informed on the subject matter to rank them in Merit Order Sequentiality. The scree plot, which displays the Eigenvalue of the components, is presented in Fig. (1).

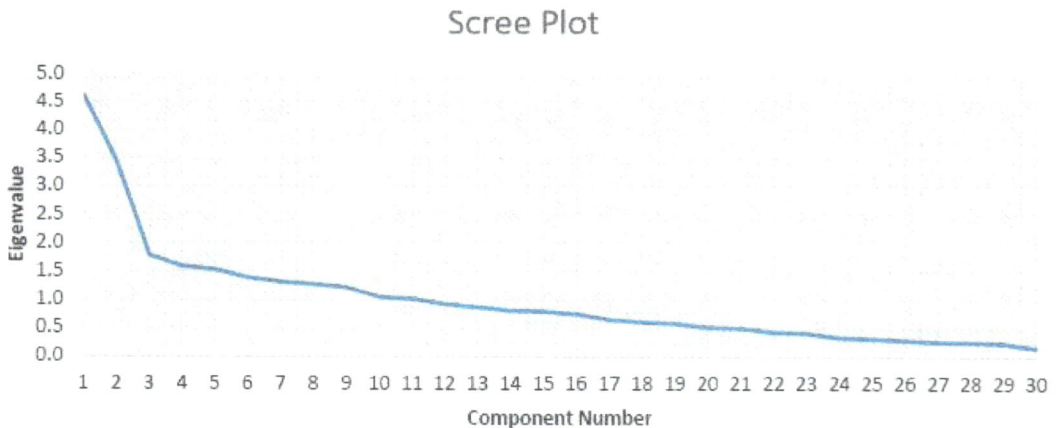

Fig. (1). Sree plot.

It is evident from the scree plot that at $\lambda = 1$ eleven factors can be extracted.

Analysis of Statistical Process Control (SPC) tools, such as Pareto, and cause-an--effect diagrams, were used to analyse customers' complaints and preferences. Minitab and Quality Management Committee (QMC) software package employed facilitated the analysis.

Fig. (**2**) depicts the Pareto chart for sixteen types of customers complaint on roofing sheets such as broken edges, weak line, and sticking sheets are classified as vital few customers' problems while the remaining thirteen were classified as trivial.

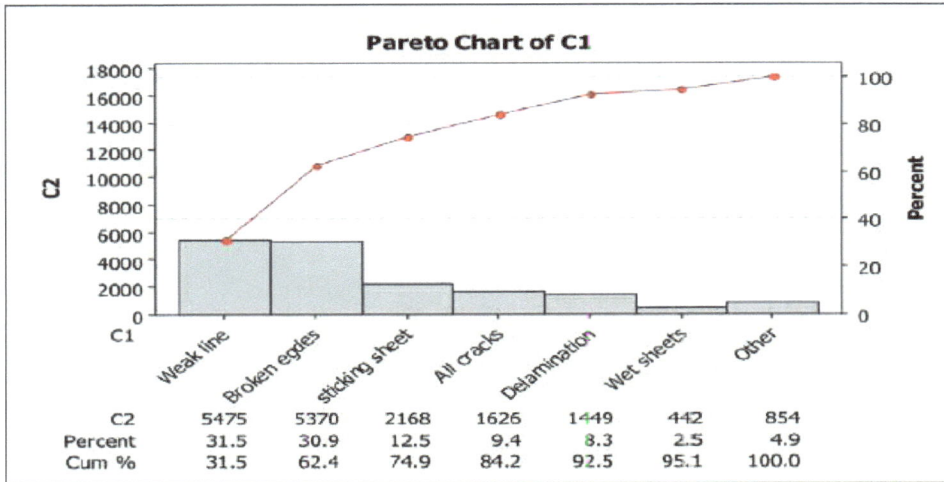

Pareto Chart of C1

C1	Weak line	Broken edges	sticking sheet	All cracks	Delamination	Wet sheets	Other
C2	5475	5370	2168	1626	1449	442	854
Percent	31.5	30.9	12.5	9.4	8.3	2.5	4.9
Cum %	31.5	62.4	74.9	84.2	92.5	95.1	100.0

Fig. (2). Pareto chart.

Fig. (**3**) and Fig. (**4**) were used to address production process and production problems, respectively in order to eliminate root causes of customers' complaints on roofing sheets.

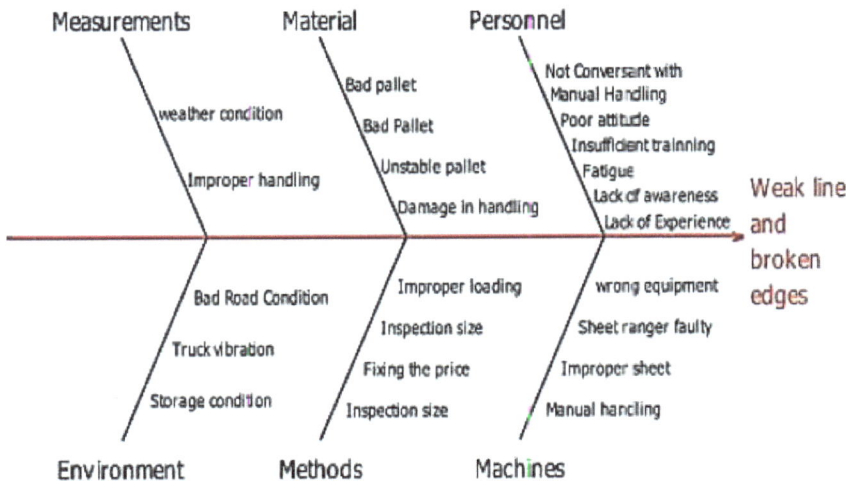

Fig. (3). Weak line and broken edges diagram.

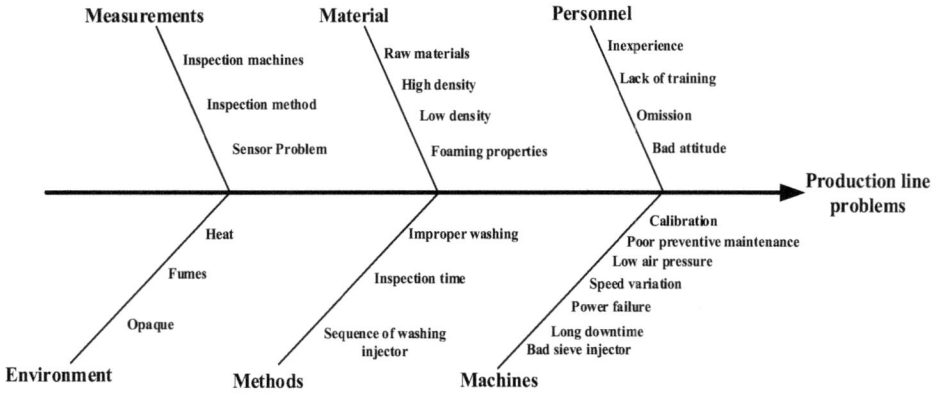

Fig. (4). Production line problems diagram.

Mean chart and R - chart

The mean and R-charts are determined as follow:

$$\bar{\bar{x}} = \frac{1}{m}\sum_{i-1}^{m}\bar{x}_i = \frac{\bar{x}_1 + \bar{x}_2 + \ldots + \bar{x}_m}{m}$$

$$\bar{X} = \frac{840.944}{600} = 1.402 \text{ Kg/dm}^3$$

Upper Control Limit =UCL $= \bar{\bar{X}} + A_2\bar{R} = 1.42 \text{ kg/dm}^3$

Lower Control Limit =LCL $= \bar{\bar{X}} + A_2\bar{R} = 1.38 \text{ kg/dm}^3$

$A_2 = 0.577, D_4 = 2.115, D_3 = 0$

Fig. (5) shows that roofing sheets production processes are within acceptable control limit for both X and R charts after implementation of QMC suggestions to address production problems that cause weak line.

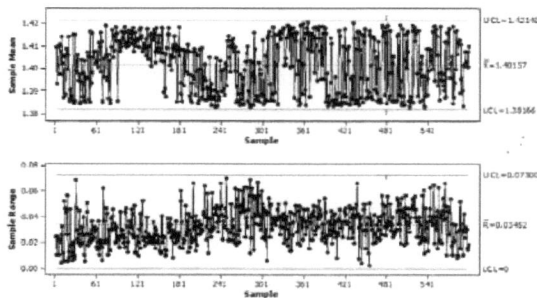

Fig. (5). X-bar and R-charts.

Range

The range is calculated thus:

$$\bar{R} = \frac{1}{m}\sum_{i=1}^{m} R_i = \frac{R_1 + R_2 - \dots + R_k}{m}$$

$$= 19.928 \div 600 = 0.033$$

$$UCL = 2.115 \times 0.033 = 0.070$$

$$LCL = 0 \times 0.033 = 0$$

Markov Chain Data Analysis for Industrial Accidents

Table **1** shows the health, safety, and environment record from fibre cement roofing sheet manufacturing company for an eleven-year (2005-2015).

Table 1. Health, safety, and environmental record.

States	2005	2006	2007	2008	Years 2009	2010	2011	2012	2013	2014	2015	Total
Fatality (FAT)	0	0	0	0	0	0	0	0	1	0	0	1
Loss Time Accident (LTA)	8	6	4	2	6	5	1	0	0	1	3	36
Medical Aid Accident (MAA)	18	15	12	6	15	12	3	5	3	4	2	95
First Aid Incident (FAI)	25	17	19	12	8	22	9	6	7	5	6	136
Near Miss (NM)	40	35	32	25	34	41	14	8	10	12	8	259
TOTAL												527

Fig. (**6**) presents the Markov chain accident transitions diagram.

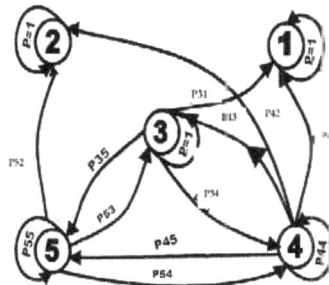

Fig. (6). Markov chain accident transitions diagram.

Fundamental Matrix Development

The from-to-matrix associated with the computed transition probability matrix (TPM) is represented as follow:

$$T = \begin{bmatrix} 1 & 0 & 0 & 0 & 0 \\ 0 & 1 & 0 & 0 & 0 \\ 0.0020 & 0 & 0.1927 & 0.2759 & 0.5254 \\ 0.0019 & 0.0681 & 0.1796 & 0.2571 & 0.4896 \\ 0 & 0.0682 & 0.1799 & 0.2578 & 0.4905 \end{bmatrix}$$

$$R = \begin{bmatrix} 0.0020 & 0 \\ 0.0019 & 0.0681 \\ 0 & 0.0682 \end{bmatrix} \quad \text{R stands for initial distribution}$$

$$Q = \begin{bmatrix} 0.1927 & 0.2759 & 0.5254 \\ 0.1796 & 0.2571 & 0.4896 \\ 0.1799 & 0.2578 & 0.4905 \end{bmatrix} \quad \text{Where Q is with transitions from non-transient to transient}$$

$$I - Q = \begin{bmatrix} 1 & 0 & 0 \\ 0 & 1 & 0 \\ 0 & 0 & 1 \end{bmatrix} - \begin{bmatrix} 0.1927 & 0.2759 & 0.5254 \\ 0.1796 & 0.2571 & 0.4896 \\ 0.1799 & 0.2578 & 0.4905 \end{bmatrix}$$

The identity matrix was regularized as shown above for ease of computations, which were carried out using MATLAB software.

.

$$N = (1 - Q)^{-1} = \begin{bmatrix} 4.2344 & 4.6327 & 8.8184 \\ 3.0142 & 5.3172 & 8.2178 \\ 3.0203 & 4.3262 & 9.2345 \end{bmatrix}$$

Where $(I-Q)^{-1}$ refers to mean habitation of the accident within the non-absorbing matrix, this is also referred to as the fundamental matrix.

$$NR = (1-Q)^{-1}R = B = \begin{bmatrix} 4.2344 & 4.6327 & 8.8184 \\ 3.0142 & 5.3172 & 8.2178 \\ 3.0203 & 4.3262 & 9.2345 \end{bmatrix} \times \begin{bmatrix} 0.0020 & 0 \\ 0.0019 & 0.0681 \\ 0 & 0.0682 \end{bmatrix} \quad B = \begin{bmatrix} 0.0773 & 0.9169 \\ 0.0161 & 0.9226 \\ 0.0943 & 0.9244 \end{bmatrix}$$

Multiple Linear Regression Analysis

Considering,

$$y = \beta_0 + \beta_1 x_1 + \beta_2 x_2 + \beta_3 x_3 + \beta_4 x_4 + \beta_5 x_5 + \beta_6 x_6 + \beta_7 x_7 + \beta_8 x_8 + \beta_9 x_9 + \beta_{10} x_{10} + \beta_{11} x_{11}$$
$$+ \beta_{12} x_{12} + \beta_{13} x_{13} + \beta_{14} x_{14} + \beta_{15} x_{15} + \varepsilon_1$$

(1)

The method of least square was used to develop a set of sixteen normal equations by converting the 16 equations into their matrix form. y is the dependent variable, while x_1 to x_{15} refers to as the independent variables.

The coefficients of the variables from the multivariate regression equations generated are transposed into matrix form. The left and right hand sides of the equations are stated as vectors A and B, respectively. The computation was accordingly done with MATLAB R2007b (7.5).

RESULTS AND DISCUSSION

In achieving the research aim and objectives, adequate measures were put in place to ensure the accuracy, reliability, and dependability of the results. Accordingly, the research, among other things, identified sixty-one scale variables that affect the fibre cement roofing sheet operations. The use of KCC and PCA as statistical tools was utilised to provide insight into the correlation among these variables.

The KCC is 0.43 which is considered adequate, suggesting that there was agreement among the judges that ranked the variables. Consequently, a null hypothesis suggesting that the ranking of variables by thirteen (13) judges dissonant was not acceptable at a p-value of 5%. The implication of re-ordering of these variables by the judges is that the problems can be hierarchically arranged in terms of management attention. The ranking of the scale items implies that management should pay more attention to the issue raised according to their severity. This will offer a veritable framework for achieving productivity in fibre cement roofing sheets.

The questionnaire couched in 5-point Rensis Likert's attitudinal was successful in extracting the responses from 143 respondent's score into a data matrix. The PCA was successful in achieving parsimony by reducing sixty-one (61) variables into thirty (30) variables.

It was obvious from the findings that customers complaint due to substandard quality products. After the implementation of QMC recommendations, weak lines and other causes of product defects were eliminated.

Thus, the multiple linear regression model developed for Fibre cement roofing sheet manufacturing data is given by the equation below:

$$\hat{y} = -394.8988 + 6.404x_1 - 8.678x_2 + 400.889x_3 + 22.590x_4 - 460.584x_5 - 0.33348x_6 + 358.704x_7 + 2.378x_8 + 1.1135x_9 + 0.1389x_{10} - 16.714x_{11} - 0.1764x_{12} + 43.612x_{13} + 3.275x_{14} - 3244.06x_{15}$$

MATRIX ERROR $= 10^{-4} \times 0.5768 = 5.768 \times 10^{-5}$

$r^2 = 1$

CONCLUSION

The study surveyed sixty-one factors that influence the performance of fibre cement roofing sheet production and established the intercorrelations among the variables. Again, the research enhances the quality output of the fibre cement-roofing sheet manufacturing using Statistical Process Control tools to solve customers' complaints. Furthermore, the research also promotes processes of fibre cement roofing sheet production, continuous improvement by use of statistical process control, development of an easy approach for the management to measure production performance by use of Correlation matrix, forecasting the output quantity of fibre sheet manufacturing using multiple linear regression analysis, and formulation and implementation of policies that will ensure the safety of staff using Markov Chain model.

CONSENT FOR PUBLICATION

Not applicable.

CONFLICT OF INTEREST

The authors declare no conflict of interest, financial or otherwise.

ACKNOWLEDGEMENTS

Declared none.

REFERENCES

[1]   Etex World [Online], https://www.etexgroup.com/en/inspiring/purpose-strategy/ [Accessed 3rd May, 2021].

[2]   A.C. Igboanugo, K.A. Bello, and C.M. Chiejine, "A factorial study of fibre cement roofing sheet manufacturing", *Journal of Multidisciplinary Engineering Science and Technology,* vol. 3, no. 2, pp. 3885-3892, 2016.

[3]   J. Balouga, *Nigerian local content: challenges and prospects.* International Association of Energy Economics, 2012, pp. 23-26.

[4]   T. Goezeglanczyk, and K. Schabowicz, "Nondestructive testing of moisture in cellulose fibre cement boards ', *11ᵗʰ European Conference on Non-destructive testing Prague Czech Republic,* 2014.

[5]   G.H.D. Tololi, R.S. Teixeira, M.A. Pereirada-Silva, F.A.R. Lohr, F.H.P. Silva, and H. Savastano, "Effects of eucalyptus pulp refining on the performance and durability of fibre cement composites", *J. Trop. For. Sci.,* vol. 25, no. 3, pp. 400-409, 2013.

[6]   R. Cechmanek, V. Pracher, and J. Loskot, "Light-weight fibre cement cladding elements", *Chemistry and Materials Research,* vol. 5, pp. 112-114, 2013.

[7]   "The Grave to the Cradle", *Global cement magazine,* pp. 12-16, 2020. [Online] Available at, http://cementassociation.ir/Library/308.pdf [Accessed 5th March, 2021].

[8]   A. Blanco, E. Fuente, L.M. Sanchez, A. Alonso, J. Tijero, and C. Negro, "Optimal use of flocculent on the manufacturer of fibre-cement materials by the Hatschek process", *10th International Inorganic-Bended Fibre –Cement Conference, Noriete,* 2006pp. 145-155.

[9]   R. Frazao, and R. Fernandes, "Comparative analysis of the life cycle of ATFibre-ement and NT fibre-cement", In: *AT Fibre-cement / NT Fibre-cement LCA – Final Report,* 2004, pp. 1-41.

[10] J. Dyczek, "Surface of Asbestos-Cement (AC) roof sheets and assessment of the risk of asbestos release. ", In: *AGH University of Science and Technology, Krakow, Poland,* 2006, pp. 28-29. In conference presentation Krackow, Poland.

[11] H.J. Lee, E.K. Park, D. Wilson, E. Tutkun, and C. Oak, "Awareness of asbestos and action plans for its exposure can help lives exposed to asbestos", *Saf. Health Work,* vol. 4, no. 2, pp. 84-86, 2013. [http://dx.doi.org/10.1016/j.shaw.2013.04.005] [PMID: 23961330]

[12] Z. Lin, "Transformation to fibre cement from asbestos cement", *International Asbestos Cement Conference,* 2014, pp. 1-28 Vienna, Austria

[13] K.A. Bello, A.C. Igboanugo, and A.C. Ovuworie, "Investigation and analysis of customers' complaints on fibre cement roofing sheet manufacturing", *International Journal of Advancements in Research & Technology,* vol. 6, no. 3, pp. 25-55, 2017.

[14] F. Sultana, N.I. Razive, and A. Azeem, "Implementation of Statistical Process Control (SPC) for manufacturing performance", *Jixie Gongcheng Xuebao,* vol. 1, pp. 15-21, 2006.

[15] O.M. Hafidz, "Statistical process control charts for measuring and monitoring temporal consistency of ratings", *J. Educ. Meas.,* vol. 47, no. 1, pp. 18-35, 2010. [http://dx.doi.org/10.1111/j.1745-3984.2009.00097.x]

[16] B.P. Mahesh, and M.S. Prabhuswamy, "Process variability reduction through Statistical Process Control for quality improvement", *Int. J. Qual. Res.,* vol. 4, no. 3, pp. 193-203, 2010.

[17] P. Besse, and J.O. Ramsay, "Principal component analysis of sampled functions", *Psychometrika,* vol. 51, pp. 285-311, 1986. [http://dx.doi.org/10.1007/BF02293986]

[18] P.D. Mishra, and J. Min, "Analyzing the relationship between dependent and independent variables in Marketing: a comparison of multiple regression with path analysis", *Innovation marketing,* vol. 6, no. 3, pp. 113-120, 2010.

CHAPTER 12

Development and Simulation of an Unmanned Aerial Vehicle (UAV)

Elizabeth Imuetiyan Omo-Irabor[1,*], **Ilesanmi Afolabi Daniyan**[2], **Adefemi Omowole Adeodu**[3] and **Titus Kehinde Olaniyi**[1]

[1] *Department of Mechanical Engineering, and Mechatronics Engineering, Afe Babalola University Ado-Ekiti, Nigeria*

[2] *Department of Industrial Engineering Tshwane University of Technology, Pretoria, South Africa*

[3] *Department of Mechanical and Industrial Engineering University of South Africa, Florida, Johannesburg, South Africa*

Abstract: The proper design of the UAV will enable the system to meet its service and functional requirements. It will also promote its availability and reliability, thereby extending its useful life in service. This work focuses on the development, sizing of an engine, and simulation of an Unmanned Aerial Vehicle (UAV), including the assembly of the whole propulsion system in the UAV. The mission of the developed UAV is for short range (25-100 km). The propulsion system consists of the following: the engine, muffler, electronic ignition system, fuel tank, engine servos, propeller, firewall, and battery. These components assembled together give the UAV its required thrust for flight. The design calculations were done, and simulations for the design were carried out using Solid Works and AutoCAD. The developed UAV was constructed using local materials such as Styrofoam and wood. It was driven by a two stroke internal combustion engine (gas engine) and an electric motor controlled by an electric speed controller (ESC). The performance evaluation of the developed UAV was carried out using the virtual X-Plane flight simulator. The results obtained indicated that that the developed UAV can be deployed for the short-range missions. This work provides design data for the development of the UAV. Hence, it is envisaged that the outcome of the study will be of immense guide to industries, which specialize in the development of UAV.

Keywords: Engine sizing, Propulsion system, Simulation, UAV.

INTRODUCTION

An Unmanned Aerial Vehicle (UAV) is defined as an aircraft without a human pilot. It is commonly known as a drone or a remote control airplane [1]. The

* **Corresponding Author Elizabeth Imuetiyan Omo-Irabor:** Department of Mechanical Engineering, and Mechatronics Engineering, Afe Babalola University Ado-Ekiti, Nigeria; Tel: +2347066125210; E-mail: tiyanomoirabor@gmail.com

Ilesanmi Afolabi Daniyan (Ed.)

UAVs are controlled either remotely or with the aid of autonomous on-board computers [2]. UAVs are mostly found in military and special operation applications, but they are also found in civil applications such as; policing, fire-fighting, surveillance, and aerial mapping [3, 4]. They are often preferred for missions that are too dirty or dangerous for manned aircraft. UAVs are small, versatile, and consume low fuel [5, 6]. The most important characteristics concerning UAV are range (how far it flies), endurance (how long it flies), and payload (what it can carry). The factors that determine the range, endurance, and payload are the power of the engine, the weight of the UAV, and efficiency of the entire system [7, 8]. UAVs can be powered with electric motors, internal combustion engines, also called gas engines (which can be in the form of wankel engines, two stroke or four stroke engines which run on glow plug and petrol) or turbine engines. The most commonly used are electric motors and internal combustion engines because of their availability, low cost, ease of installation, and maintenance [9, 10]. Although electric motors are cheaper to buy and maintain, less noisy, and easier to install when compared to gas engines, for the purpose of this work, gas engines and turbines will be considered. Gas engines are preferred because they have a high power to size ratio and are much more environmental and user friendly [11, 12]. Turbine engines, on the other hand, although are more expensive, produce the most amount of power when compared to all other types of engines [13,14]. According to Dix [15] and Uzol [16], UAVs typically fall into one of six functional categories (although multi-role airframe platforms are becoming more prevalent), namely target and decoy (providing ground and aerial gunnery a target that stimulates an enemy aircraft or missile), reconnaissance (providing battlefield intelligence), combat (providing attack capability for high risk missions), logistics (UAVs specifically designed for cargo and logistics operation), research and development (used to further develop UAV technologies to be integrated into field-deployed UAV aircraft) as well as civil and commercial (specifically designed for civil and commercial applications). The propulsion technologies are various technologies that are able to provide the thrust that enables a UAV to move. Technologies used in powering a UAV include Internal Combustion Engines (ICE), turbine engines, electric motors, *etc.* Moreover, ICE and electric motors are common due to availability, low cost, ease of installation, and maintenance. Many researchers have worked on the engine sizing and simulation of an Unmanned Aerial Vehicle (UAV) [17 - 20], but this work improves the existing UAV in the areas of structural design, stability, flight capability, and manufacturing process as well as process optimization. The aim of the work is to carry out a functional, cost effective engineering design and construct a UAV with a focus on preliminary engine design, which would provide enough power to turn the propeller and produce sufficient thrust for continuous flight. The UAV is specifically developed for surveillance whose endurance can last up to 1 hour. Many countries are ill equipped in the areas of intelligence gathering and surveillance missions over suspected or discovered trouble spots. This is not only due to the lack of functional aircraft for such roles but also the high cost associated with flying aircrafts on such long runs. Also, agricultural organisations, geographical mapping agencies, and departments lack aerial coverage and data because of the lack of satellite surveillance data provided for them. The need for the development of Unmanned Aerial Vehicles is therefore very important to help solve these problems in developing countries. Beyond the military applications of Unmanned Aerial Vehicle with which they are most associated, numerous civil aviation uses have been developed, including aerial surveying of crops, acrobatic aerial footage in filmmaking, search and rescue operations, inspecting power lines and pipelines, counting wildlife, delivering medical supplies to remote or otherwise inaccessible regions, scientific research, *etc.* [21]. Hence, the design and construction of a surveillance UAV are of immense importance in the local and international society due to its numerous advantages. For instance, a well-designed UAV system can meet service and functional requirements.It will also promote its availability and reliability, thereby extending its useful life in service. The novelty of this work lies in the fact that the preliminary engine sizing and simulation analysis of UAVs using virtual X-Plane flight simulators for short-range missions has not been sufficiently highlighted by existing literature. This ensures a better understanding of the operations of the UAV. The design, construction, and simulation of UAVs are of immense importance in the local and international society as well as the aerospace industry for surveillance purposes.

MATERIALS AND METHOD

The design calculations and simulations for the design were done using Solid Works, AutoCAD, and X Plane Simulator, and then the parts for the UAV's construction were ordered.The UAV developed was constructed using local materials such as styrofoam and wood. It was driven by a two stroke internal combustion engine (gas engine), an electric motor controlled by an electric speed controller (ESC) (Fig. **1**).

Fig. (1). The developed UAV.

The propulsion system consists of the engine, muffler, electronic ignition system, fuel tank, engine servos, propeller, firewall, and battery. These components assembled together give the UAV its required thrust for flight.

ENGINE SIZING OF A UAV

This section describes the scaling of an engine to provide the thrust necessary to overcome drag based on the different types of mission requirements presented in Table 1. The mission of the developed UAV is for short range (25-100 km); according to Beard and McLain [22], the critical parameters for the short-range mission include the altitude, speed, and thrust to weight ratio.

Table 1. Thrust-to-weight ratio based on mission requirement for different historic aircraft [7].

S/N	Primary mission requirement	T/W
1.	Long Range	0.20-0.35
2.	Short and Intermediate Range	0.30-0.50
3.	STOL	0.40-0.60
4.	Combat: Close-Air Support	0.40-0.60
5.	Combat: Air-to-Air	0.80-1.30
6.	Combat: High-Speed Intercept	0.55-0.80

The UAV under consideration is in the short and intermediate range. The thrust to weight ratio should fall between 0.30-0.50 [20]. The expected weight of the UAV is 5 N. Since the weight and T/W ratio are known, the expected performance of the engine can be determined from Equation (1).

$$T/W = \frac{Thrust\ of\ Engine}{Weight\ of\ UAV} \tag{1}$$

For a weight of 5 N using the maximum T/W, which is 0.50.

$$0.5 = \frac{X}{5}; \quad X = 2.5$$

Therefore, the expected performance of the engine should be at most 2.5-horse power (HP).

The maximum gross weight of the aircraft (W_0) is the sum of the aircraft empty weight (W_e), total payload weight (W_p), total fuel weight (W_f), and miscellaneous weight (W_{misc}) expressed as Equation (2).

$$W_0 = W_e + W_p + W_f + W_{misc} \tag{2}$$

The aircraft empty weight (W_e) is expressed by Equation (3).

$$W_e = W_a + W_L + W_{eng} + W_s \tag{3}$$

Where: W_a is the weight of the airframe (kg), W_l is the weight of the landing gear (kg), W_{eng} is the engine weight (kg) and $W_s$is the weight of the system (kg).

The range of the aircraft is expressed by Equation (4).

$$R = R_{CL} + R_{CR} + R_D \tag{4}$$

where: R is the range (km), R_{CL} is the climb distance (km), is the cruise range (km)R_{CR} and R_D is the descent distance (km).

The flight time (t_f) is expressed as Equation (5).

$$t_f = t_{CL} + t_{CR} + t_D \tag{5}$$

Where: t_{CL} is the climb time (hr), t_{CR} is the cruise time (hr) and t_D is the descent time (hr)

The cruise range $R_{CR}$and the cruise time (t_{CR}) is expressed by Equations (6 and 7), respectively.

$$R_{CR} = \frac{V_0}{C}\frac{L}{D}\ln\frac{w_i-1}{w_i} \tag{6}$$

Where: V_0 is the specific fuel consumption [kg/(kN.hr)], LD is the lift to drag ratio, wi is the engine air flow (kg/sec).

$$t_{CR} = \frac{L/D}{C}\ln\frac{w_i-1}{w_i} \tag{7}$$

The power coefficient of the propeller is expressed as Equation (8).

$$C_p = \frac{P_s}{n^3 d^5 \rho} \tag{8}$$

where: C_p is the power coefficient of the propeller, P_s is the uninstalled engine shaft power (kW), n is the propeller speed (rev/sec), d is the propeller diameter (m), and p is the air density (kg/m^3)

Going through the Data Life Engine (DLE) Catalogue [23], the most suitable engine is the DLE 20cc Gasoline Engine. The specifications of the engine are as follows: Performance: 2.5HP/ 9000 rpm, Idle Speed: 1,700 rpm, Ignition Style: Electronic Ignition, Recommended Propellers: 14A-10, 15A-8, 16A-6, 16A-8 and 17A-6, Diameter A- Stroke: 32mm A- 25mm, Compression Ratio: 10.5:1, Fuel: Gasoline with a 0:1 gas/2 stroke oil mixture, Weight: Main Engine: 650 g, Muffler: 50 g and Electronic Ignition:120 g.

Engine Ratings

According to Jenkinson *et al.* [13], the engine rating specifies the performance of an engine under various conditions. These ratings correspond to different thrust conditions that are specified for take-off, maximum climb, and maximum cruise, so it is important to take note of them.

Take-Off

The take-off rating is the maximum thrust that the engine is certified to produce. This is generally specified for short periods of time, of the order of five minutes, to be used only at take-off. The take-off rating is generally used when "sizing" an engine for a design. The maximum climb rating is the maximum thrust that the engine is certified to produce for normal climb operation. This rating is from 90-93% of the take-off rating. The maximum cruise rating is the maximum thrust that the engine is certified to produce for a normal cruise. This corresponds to 80% of the take-off rating. Also, the cruise rating is for continuous operation, with no time limits.

Components of the Propulsion System

The design consideration of the propulsion system takes into account the short-range mission of the UAV. The various components of the propulsion system and their functions are listed in Table **2** and shown in Figs. (**2 - 9**).

Table 2. Components of the propulsion system and their functions.

S.NO.	Component	Function
1.	Propeller	Converts the rotary motion of the engine into thrust to lift the UAV
2.	Engine	Produces the rotary motion that powers the propeller
3.	Electronic Ignition	Provides the ignition for the engine
4.	Engine Mount	Provides rigid support for the engine to rest on

(Table 2) cont.....

S.NO.	Component	Function
5.	Firewall	Protects the engine from other propulsion components in the UAV
6.	Engine Servos	There are two: the choke and throttle servo. They connect the choke and throttle from the engine to the transmitter.
7.	Battery	It powers the electronic ignition
8.	Fuel Tank	It holds the fuel for the engine
9.	Fuel Line	It conveys fuel from the tank to the engine
10.	Fuel Filter	It helps to prevent dirt and other impurities from entering the engine

Fig. (2). Engine and muffler.

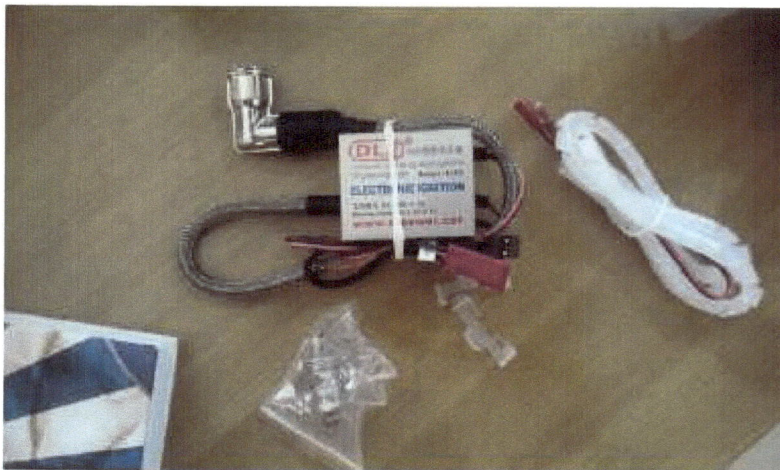

Fig. (3). Engine mount.

Fig. (4). Engine mount.

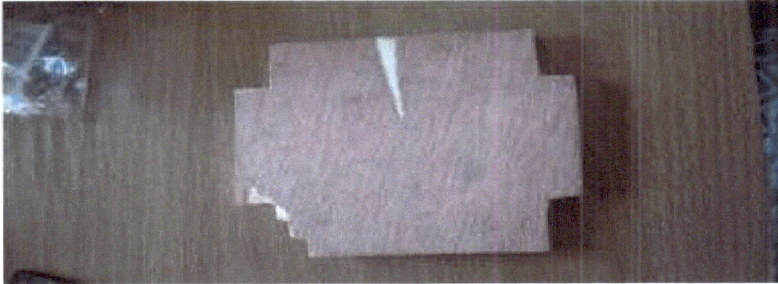

Fig. (5). Firewall.

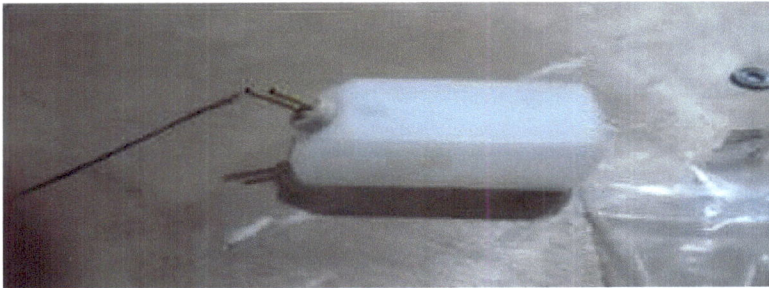

Fig. (6). Fuel tank.

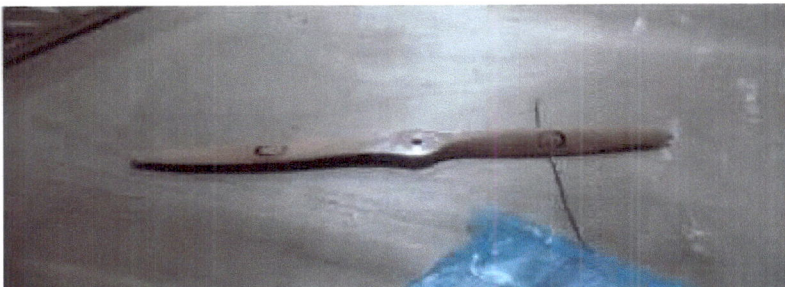

Fig. (7). 17×8 Propeller.

Fig. (8). Battery.

Installation of the Propulsion System

The various steps taken in the installation of the propulsion system for the UAV are explained in detail below.

The firewall was made of two separate 6 mm thick plywood strips (Fig. **5**). The two strips were 127 mm A- 127 mm in size. The two strips were joined using a solvent-based adhesive (Top Bond). The plywood strips were arranged in a cross grain pattern to enhance strength and load transference.

Using the DLE 20 manual [23], the points for the assembling the engine mount were marked, and holes were drilled using a drilling machine. The engine mount was then assembled on the firewall using nuts and bolts. Then a portion of the firewall was cut to allow the connection of the servos to the engine. The engine mount alone and assembled on the firewall is shown in Fig. (**10** and **11**), respectively.

Fig. (10). Engine mount as standalone.

Fig. (11). Engine mount (assembled on the firewall).

The engine was placed on the engine mount at a distance from the firewall to test-fit the position (Fig. **12**). When the satisfactory position was gotten, the locations for the holes on the mount were marked and drilled. Then the engine was secured to the mount using bolts and nuts. The engine was assembled upside down due to the servo placement and also for aesthetic value.

Fig. (12). Assembled engine on the mount.

The first pair of accessories to be installed is the engine servos for the choke and throttle. They are installed behind the engine and connected to the engine with the aid of ball links, pushrods, and a screw lock connector so that the linkage is adjustable. The next accessory to be installed is the fuel tank and fuel lines. A strip of plywood is glued on top of the servos at a reasonable distance. Holes are then drilled on the plywood, and two rubber bands are used to secure the fuel tank. The fuel tank has two lines, one for the vent and the other is connected to the carburettor. Then the electronic ignition is installed. The main wire is first covered with silicon wire to prevent friction, and then the ignition module is securely placed behind the engine (Fig. **13**).

Fig. (13). Assembled engine accessories.

During the installation of the propeller, the first component to be installed is the spinner backplate, then the propeller, the propeller washer, the adapter nut, and finally the spinner cone are installed on the crankshaft (Fig. **14**).

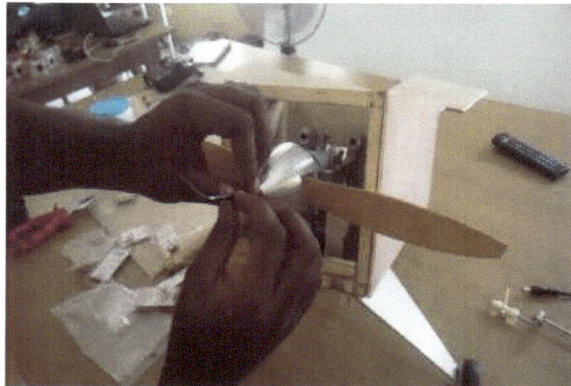

Fig. (14). Installation of the propeller.

Engine Testing

The procedural steps for starting the engine are stated as follow:

  i. The battery was connected to the electronic ignition, and the electronic ignition was connected to the engine.
 ii. The propeller was installed at the one o'clock position and at the beginning of the compression stroke so that it is comfortable to flip it through compression.
iii. A team member helped to hold the plane while the engine was being started.
 iv. The ignition was switched on, the choke on the carburettor was closed, and

the throttle was opened slightly from the idle position.

v. The propeller was flipped briskly several times until a popping sound was heard; this indicated that the engine was firing.

vi. The choke lever was moved to the open position.

vii. The throttle was set to a high idle, and the propeller was set so that it was at the beginning of the compression stroke.

viii. The propeller was flipped through compression rapidly. According to the engine manual, if this is done properly, the engine will start after several brisk flips of the propeller.

ix. According to the engine manual, if your engine did not start, repeat steps 3-8.

The Surveillance System

The surveillance system is made of two-camera system; the first camera is a high definition camera for surveillance, while the second camera is a Charge-Coupled Device (CCD) board camera solely for piloting. The first and the second cameras have a pan and tilt system, with the second placed at a 45° angle to the ground. The entire module did not exceed a total weight of 0.1 kg. The system is capable of transmitting video feeds before, during, and after the flight. The integration and system-testing phase followed after the implementation and unit testing, with the aim of bringing all of the different units together to form a system.

Virtual Test Flight and Simulation

The performance evaluation of the developed UAV was carried out using the virtual X-Plane flight simulator. The software is suitable for aerodynamics modelling, simulation, and calculations, as well as in analysis of flight dynamics and control. The physical parameter of the UAV that serves as the input parameters in the X-Plane flight simulator are as follow; total weight of UAV (5 kg); thrust to weight ratio (0.50); thrust of engine (2.5 hp); wingspan (1.450 m); body length (1.2 m); wing reference area (0.354 m²); wing sweep angle (0.0°); pitch moment of inertia (2.687 kg.m²); mean chord (0.2 m); angle of attack (2.98°); lift coefficient (0.5); drag coefficient (0.0247); altitude (20-100 km); throttle (48%); pitch moment coefficient (0.00295).

RESULTS AND DISCUSSION

Fig. (15) is a plot of altitude and time. The UAV is designed for short-range missions. The maximum altitude reached was 150 m in 5 minutes at a speed of 0.05 m/s. The altitude is later observed to decrease as the time of flight increases. The total time of flight was 14 minutes. The engine rotational frequency, speed, and weight of the UAV are critical factors determining flight time.

Altitude (km) vs time (min)

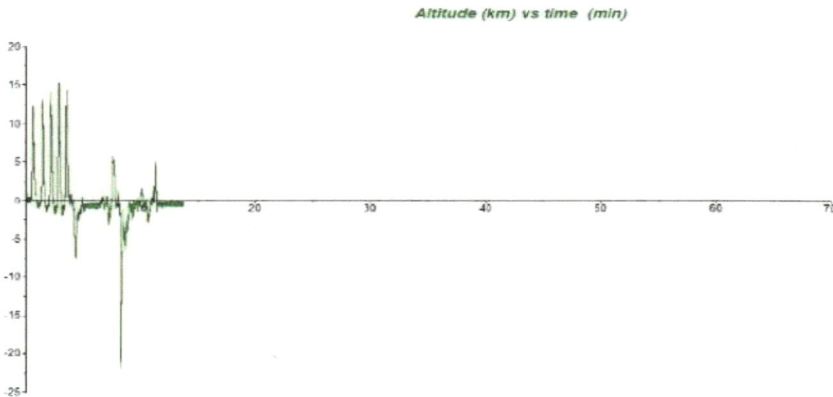

Fig. (15). Plot of altitude and time.

From (Fig. **16**), the maximum pitch angle was 3.8° within the total time of flight of 14 min. The vertical force applied at a forward distance from the centre of the UAV causes it to pitch up and down at a maximum angle of 3.8°. This is not likely to offset the balance of the UAV as the pitch angle can vary between 0°-25° [22]. An inner loop stability control can be employed to achieve the desired pitch angle.

Pitch angle (deg) vs time (min)

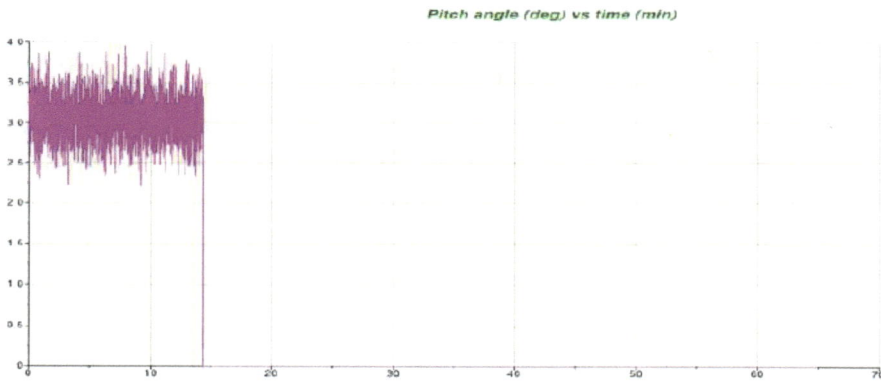

Fig. (16). Plot of pitch angle and time.

Fig. (**17**) shows the variation of the engine's rotational frequency with time. At a total time of flight of 14 min, the rotational frequency was 285 rpm. The power output of the UAV is primarily a function of the engine's speed.

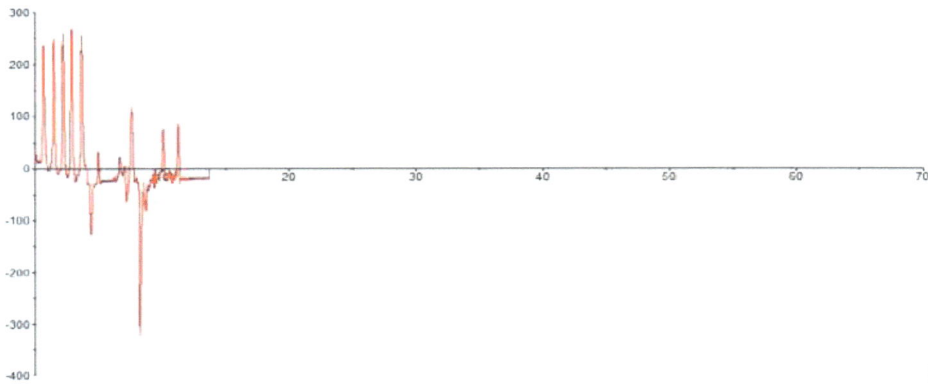

Fig. (17). Plot of rotational speed and time.

The maximum throttle position was 48% (Fig. **18**). If the UAV is to produce a continuous flight in the air with a half throttle, then the rotor must produce about 50% higher thrust as compared to the multi-rotor.

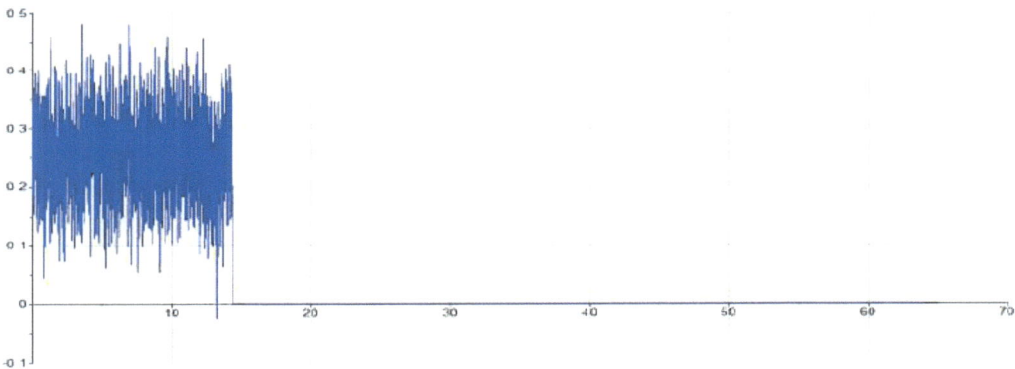

Fig. (18). Plot of the throttle position and time.

Fig. (**19**) shows the variation of the value of torque with speed. The maximum torque generated was 32.67 Nm. The torque determines the responsiveness of the UAV to changes in the air. An increase in the value of the torque increases the responsiveness of the UAV to changes in the air with the corresponding decrease in the speed and *vice versa;* hence, increasing the value of the toque at low speed is desirable for this operation. For high performance mission, the operation of the engine can be kept within the maximum torque-speed range.

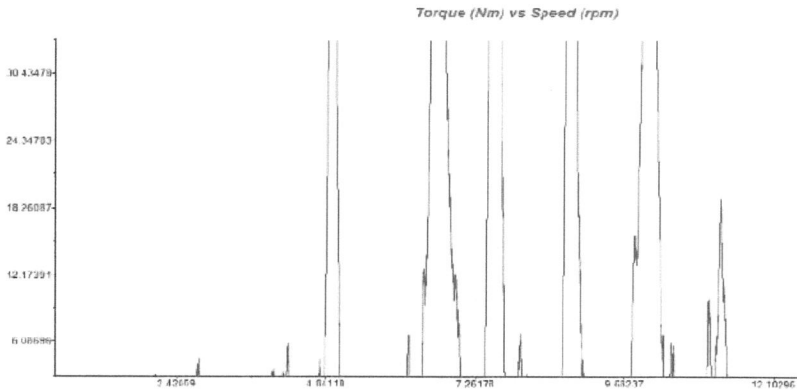

Fig. (19). Plot of torque and speed.

After all the tests were successfully carried out, the system was allowed to operate for 1 hr for performance evaluation. The system was able to provide an uninterrupted video feed, and the pan and tilt system worked as designed. The board cameras gave video feeds in real time with little or no time delay, unlike the high definition cameras, which have more time delay due to the number of feeds it has to process because of a higher amount of pixels.

CONCLUSION

The development of UAVs developed using local materials was successfully carried out. The performance evaluation of the developed UAV was carried out using the virtual X-Plane flight simulator. The simulation results obtained indicated that the developed UAV could be deployed for short range missions. This work provides design data for future scaling of its development, and the simulation results are used in the aerospace industry todetermine critical parameters of UAV during the developmental process. The application of the design parameters has the potential of enhancing the development of UAVs that will enable the system to meet its service and functional requirements.

CONSENT FOR PUBLICATION

Not applicable.

CONFLICT OF INTEREST

The authors declare no conflict of interest, financial or otherwise.

ACKNOWLEDGEMENTS

Declared none.

REFERENCES

[1]   Ö Turan, İ Orhan, and T.H Karakoç, "On-design analysis of high bypass turbofan engines", *Journal of Aeronautics and Space Technologies,* vol. 3, no. 3, pp. 1-8, 2008.

[2]   T. Olaniyi, "Design of an Unmanned Aerial Vehicle (UCAV): A case study for a developing nigerian economy", In: *MITT iSTREAMS Conference, Ado Ekiti, Nigeria,* 2015, pp. 353-360.

[3]   B. Garner, *Model Airplanes Propeller,* 2009, pp. 1-24. [Online] Available at, https://www3.nd.edu/~ame40462/garner_Model_Propellers_Article_2009.pdf [Accessed 17th June, 2020].

[4]   R.R. Glassock, Design, modelling and measurement of hybrid power-plant for Unmanned Aerial Vehicles (UAVs)., 2012.

[5]   E. Eastop, and T. McConkey, "Applied Thermodynamics for Engineering Technologists", In: *Delhi: Pearson Education*New Delhi, 2009.

[6]   A. Dinc, "Sizing of a turboprop unmanned air vehicle and its propulsion system", *Journal of Thermal Science and Technology,* vol. 35, no. 2, pp. 53-62, 2015.

[7]   T. Corke, "Design of Aircraft", In: *New Jersey: Pearson Education* Pearson Education: US, 2002.

[8]   H. Curtis, A. Filippone, M. Cook, L.R. Jenkinson, and F. De Florio, *Aerospace Engineering Desk Reference* 1st ed. Butterworth-Heinemann: San Diego, CA, 2009, p. 7.

[9]   M. İlbaş, and M. TA1/4rkmen, "Estimation of Exhaust Gas Temperature Using Artificial Neural Network in Turbofan Engines", *Journal of Thermal Science and Techn.,* vol. 32, no. 2, pp. 11-18, 2012.

[10]   T. Olaniyi, and O. Fadare, "Development of an Unmanned Aerial Vehicle (UAV): Focusing on empennage, wings and mechatronics (controls)", In: *Multidisciplinary Innovations and Technology Transfer (MITT) iSTREAMS* Conference: Ado-Ekiti, Nigeria, 2015, pp. 249-258.

[11]   Z. Şahin, M. Kopaç, and N.Ö Aydin, "The investigation of increasing of the efficiency in the power plant with gas-solid fuels by exergy analysis", *Journal of Thermal Science and Technology,* vol. 31, no. 1, pp. 85-107, 2011.

[12]   I. Arsie, A. Cricchio, and C. Pianese, "A comprehensive powertrain model to evaluate the benefits of Electric Turbo Compound (ETC) in reducing CO2 emissions from small diesel passenger cars", *SAE Technical Paper,* pp. 1-10, 2014.

[13]   L. Jenkinson, P. Simpkin, and D. Rhodes, ""Civil Jet Aircraft Design"", In: *Oxford: Butterworth Heinemann* Butterworth Heinemann: UK, 1999.
[http://dx.doi.org/10.2514/4.473500]

[14]   T. Olaniyi, and O. Odetayo, "Development of unmanned aerial vehicle in an emerging nigerian economy", *MINTT Conference iSTEAMS,* 2015, pp. 301-306 Ado-Ekiti, Nigeria.

[15]   D.M. Dix, "Development of engines for unmanned air vehicles: Some factors to be considered", *DA Document D-2788, Institute for Defense Analyses, 4850 Mark Center Drive, Alexandria, Virginia,* pp. 1-91, 2003.

[16]   O. Uzol, "A new high-fidelity transient aerothermal model for real-time simulations of the T700 helicopter turboshaft engine", *J. of Ther. Sci. & Tech.,* vol. 31, no. 1, pp. 37-44, 2011.

[17]   S. Chiesa, S. Farfaglia, and N. Viola, "Design of all electric secondary power system for future advanced MALE UAV", *Proc. Inst. Mech. Eng. Part G J. Aerosp. Eng.,* vol. 226, pp. 1255-1270, 2012.
[http://dx.doi.org/10.1177/0954410011420914]

[18]   T. Donateo, M. De Giorgi, A. Ficarella, E. Argentieri, and E. Rizzo, "A general platform for the modeling and optimization of conventional and more electric aircraft", *SAE Technical Paper,* pp. 1-11, 2014.
[http://dx.doi.org/10.4271/2014-01-2187]

[19]   T. Donateoa, L. Spedicatoa, G. Trulloa, A.P. Carluccia, and A. Ficarella, "Sizing and simulation of a piston-prop UAV. ATI 2015 - 70th Conference of the ATI Engineering Association", *Energy Procedia,* vol. 82, pp. 119-124, 2015.

[20]   A.P. Carlucci, A. Ficarella, and D. Laforgia, *Energy Convers. Manage,* vol. 101, pp. 470-480, 2015.
        [http://dx.doi.org/10.1016/j.enconman.2015.06.009]

[21]   W.R. Silva, A.L. Silva, and H.A. Grundling, "Modelling, simulation and control of fixed-wing Unmanned Aerial Vehicle (UAV)", *24ᵗʰ ABCM International Congress of Mechanical Engineering,* 2017pp. 1-11 Curitiba, PR, Brazil.
        [http://dx.doi.org/10.26678/ABCM.COBEM2017.COB17-2703]

[22]   R. Beard, and T. McLain, *Small-Unmanned Aircraft: Theory and Practice.* Princeton University Press. ISBN9780691149219, 2012.
        [http://dx.doi.org/10.1515/9781400840601]

[23]   "Data Life Engine (DLE) 200 Operator's Manual", [Online] Available at, http://www.dleguide.com/ [Accessed 24th June, 2020].

SUBJECT INDEX

W

Z

www.ingramcontent.com/pod-product-compliance
Lightning Source LLC
Chambersburg PA
CBHW050846220326
41598CB00006B/445